W9-CCZ-630

LAND *of* WONDROUS COLD

The Race to Discover Antarctica and Unlock the Secrets of Its Ice

Gillen D'Arcy Wood

Princeton University Press
Princeton & Oxford

Published by Princeton University Press
41 William Street, Princeton, New Jersey 08540
6 Oxford Street, Woodstock, Oxfordshire OX20 1TR

press.princeton.edu

Library of Congress Cataloging-in-Publication Data

Names: Wood, Gillen D'Arcy, author.
Title: Land of wondrous cold : the race to discover Antarctica and unlock the secrets of its ice / Gillen D'Arcy Wood.
Description: Princeton, New Jersey : Princeton University Press, 2020. | Includes bibliographical references and index.
Identifiers: LCCN 2019029836 (print) | LCCN 2019029837 (ebook) | ISBN 9780691172200 (hardback) | ISBN 9780691201689 (ebook)
Subjects: LCSH: Antarctica—Discovery and exploration. | Antarctica—Environmental conditions. | Ice caps—Antarctica.
Classification: LCC G870 .W66 2020 (print) | LCC G870 (ebook) | DDC 919.89—dc23
LC record available at https://lccn.loc.gov/2019029836
LC ebook record available at https://lccn.loc.gov/2019029837

British Library Cataloging-in-Publication Data is available

Editorial: Ingrid Gnerlich and Arthur Werneck
Production Editorial: Mark Bellis
Text and Jacket Design: Pamela Schnitter
Production: Jacqueline Poirier
Publicity: Sara Henning-Stout and Katie Lewis

Jacket Credit: The *Erebus* and *Terror* against the Ross Ice Shelf in the Ross Sea (detail), from James Clark Ross, *A Voyage of Discovery*, 1847

This book has been composed in Garamond Premier Pro and Bodoni 72

Printed on acid-free paper. ∞

Printed in the United States of America

10 9 8 7 6 5 4 3 2 1

To my friends and family in the Southern Hemisphere

—a kind of homecoming

And now there came both mist and snow,
And it grew wondrous cold;
And ice, mast high, came floating by,
As green as emerald.

—*Coleridge*

CONTENTS

PART THREE ‡ TRIUMPH

ILLUSTRATIONS

TIMELINE OF ANTARCTIC EXPLORATION, 1772–1917

From James Cook to the Heroic Age

Land of Wondrous Cold tells the story of the pioneer Antarctic voyages of 1838–1842, when British, French, and American commanders raced each other to the South Pole. As the first major scientific research expeditions in Antarctica, these early Victorian-era explorers laid the foundation for our modern understanding of the white continent, its glacial history, and the future of its all-important ice cap.

- 1772 Yves-Joseph Kerguelen sights "Desolation Island" in the sub-Antarctic waters of the Indian Ocean.
- 1773 James Cook makes the first crossing of the Antarctic Circle; he turns back upon reaching the ice pack at 67°15′ south.
- 1774 Cook achieves a record southing of 71°10′, off the coast of West Antarctica.
- 1820 Gottlieb von Bellingshausen, leading a Russian-financed expedition, sails within twenty miles of the Fimbul ice shelf, the first recorded sighting of the Antarctic continent.
- 1823 James Weddell, sailing poleward from the Atlantic South Shetland Islands, makes a new record southing of 74°15′.

1832 Samuel Enderby, the whaling magnate, finances an expedition led by John Biscoe, who makes sighting of the northern Antarctic Peninsula, now called Graham Land.

1836 The United States Congress approves funding for a large-scale exploring expedition, to include a mission of discovery to the South Pole.

1837

January Pacific explorer Dumont D'Urville proposes a third southern voyage to King Louis-Philippe of France. An Antarctic campaign is included in his orders.

September The French ships *Astrolabe* and *Zélée* sail from Toulon.

1838

January D'Urville's first Antarctic campaign, in the Weddell Sea, is thwarted by pack ice.

August The United States Exploring Expedition, commanded by Charles Wilkes, leaves Norfolk, Virginia.

1839

February The British Prime Minister, Lord Melbourne, approves a British Antarctic Expedition, to be commanded by James Clark Ross; sealer John Balleny, sailing south of Tasmania, glimpses the East Antarctic coast at 65° south.

March The US Exploring Expedition mounts its first Antarctic campaign. The schooner *Flying Fish* nears Cook's record southing in West Antarctica.

September The British polar ships *Erebus* and *Terror* set sail from Margate.

December The Wilkes expedition departs Sydney for the Antarctic; D'Urville's ships sail southward from Hobart.

1840

January The American and French expeditions, which briefly encounter each other, explore the East Antarctic coast. The French make landfall and raise the tricolor flag. Wilkes charts 1,500 miles of coast.

March British Antarctic expedition explores Kerguelen Island.

November The *Astrolabe* and *Zélée* arrive back in France; Ross sails toward the pole from Hobart.

December President Van Buren announces the American discovery of Antarctica in his State of the Union address.

1841

January The British expedition explores the Ross Sea, posts record southing in the waters off Mount Erebus, at 78°9′ south.

December First volume of D'Urville's Antarctic voyage narrative is published in Paris; Ross returns to Antarctica, but is unable to better his first attempt. Antarctic exploration enters a long period of hiatus.

1898 Norwegian Carsten Borchgrevink, first of the "Heroic Age" explorers, is the first to return to 78° south since James Ross. He overwinters at Cape Adare.

1901–1904 Robert Scott's first expedition, aboard *Discovery*, retraces Ross's route and makes land exploration to 82°17′ south.

1901–1903 Swedish expedition, led by Otto Nordenskjöld, spends two winters on the Antarctic Peninsula, with outstanding scientific results.

1907 Ernest Shackleton's first expedition, aboard *Nimrod*, makes an attempt on the South Pole, but turns back ninety-seven miles short. The expedition's northern party, including Australian Douglas Mawson, is first to ascend Mount Erebus and reach the South Magnetic Pole.

1910 Race to the South Pole between Norwegian Roald Amundsen and Englishman Robert Scott. Amundsen arrives first, on December 14, 1911. A month later, Scott reaches the pole, but he and four companions perish on the return journey.

1911 Douglas Mawson's Australian expedition, the first to follow in the wake of Dumont D'Urville and Charles Wilkes, explores the East Antarctic coast.

1914 Shackleton's attempt to traverse the Antarctic Continent goes awry when the *Endurance* becomes trapped and sinks in the Weddell Sea ice pack. Shackleton's men survive, but his Ross Sea party, laying stores for a journey that never eventuated, lose three men before their rescue in January 1917. The Heroic Age of Antarctic exploration concludes.

A NOTE ON MEASUREMENTS

Shifts in the text between imperial and metric systems reflect the two cultures, and two time periods, that are the book's focus. I have used imperial measures, a Victorian vocabulary, when writing from the point of view of the nineteenth-century explorers, and as a default. I use metric when drawing from modern scientific literature on Antarctica.

LAND *of* WONDROUS COLD

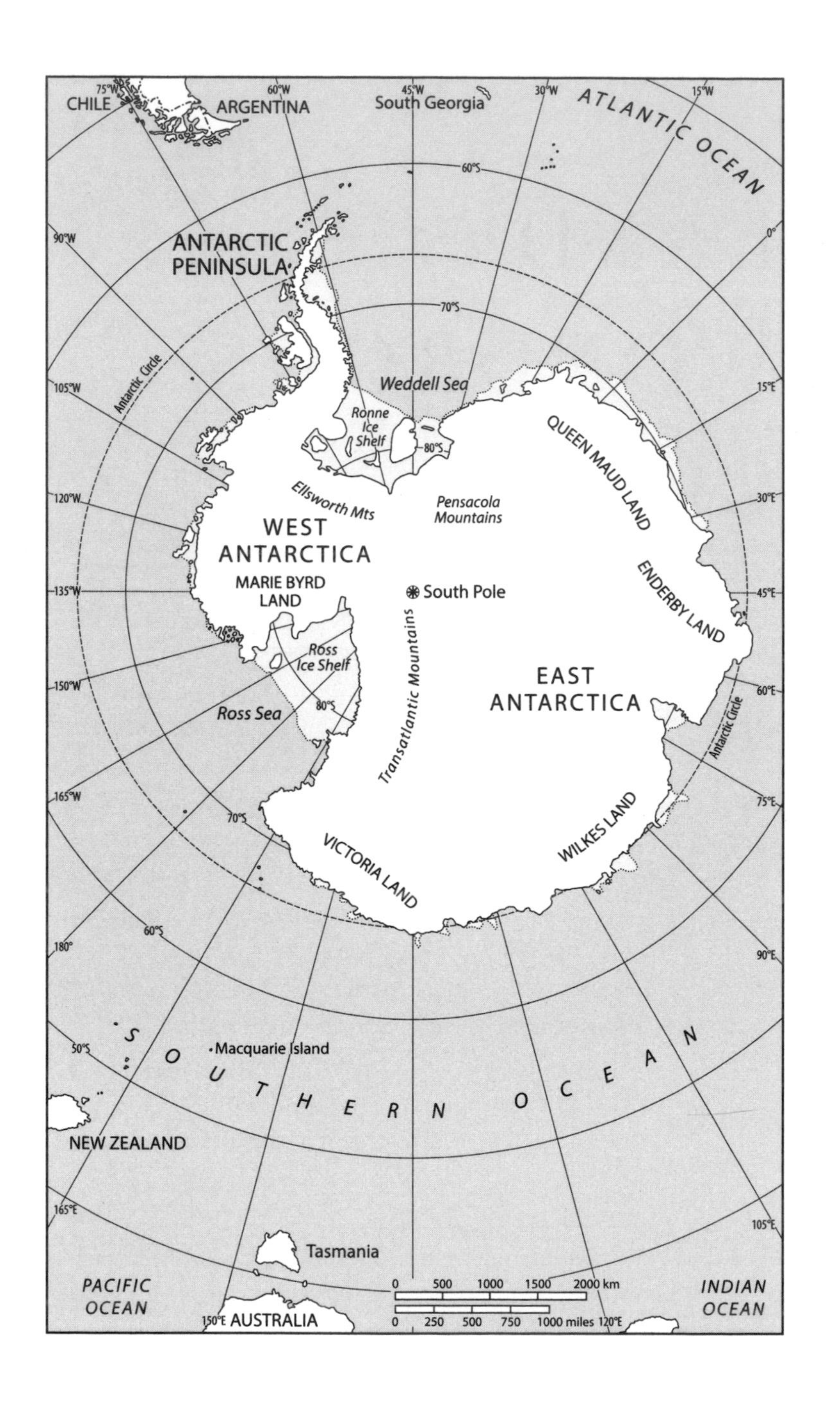
CHILE
ARGENTINA
South Georgia
ATLANTIC OCEAN
ANTARCTIC PENINSULA
Antarctic Circle
Weddell Sea
Ronne Ice Shelf
QUEEN MAUD LAND
Ellsworth Mts
Pensacola Mountains
WEST ANTARCTICA
MARIE BYRD LAND
South Pole
ENDERBY LAND
Transatlantic Mountains
Ross Ice Shelf
EAST ANTARCTICA
Ross Sea
WILKES LAND
VICTORIA LAND
SOUTHERN OCEAN
Macquarie Island
NEW ZEALAND
Tasmania
PACIFIC OCEAN
AUSTRALIA
INDIAN OCEAN
0 500 1000 1500 2000 km
0 250 500 750 1000 miles

INTRODUCTION

Our Glacial Earth

Forty thousand tourists visit Antarctica each year. Millions more have seen nature documentaries about the frozen continent and its charismatic wildlife: penguins, seals, and whales abroad on their favorite feeding grounds. Every southern summer—from December through February—hundreds of scientists descend on Antarctica armed with research grants. Over eighty research stations, accessed by air and sea, are scattered across a continent the size of the United States and Mexico combined. These stations support scientific studies of all kinds: from polar geology and glaciology to marine microbiology and paleoclimatology. Antarctica is remote from all human habitation, but not from human consciousness and endeavor.

For all the recent interest in Antarctica, myths and misperceptions persist. For example, many people assume that Terra Australis Incognita was discovered in the early twentieth

---

Fig. I.1. Mountainous East Antarctica forms the bulk of the continent, while West Antarctica and the Antarctic Peninsula are lower-lying and more vulnerable to glacial melt.

century—during the so-called Heroic Age of polar exploration—with the British *Endurance* expedition, and the tragic deaths of Captain Robert Scott and his men on their return from the South Pole, having been beaten to the prize by Roald Amundsen. But, in fact, the discovery of the Antarctic continent was achieved seventy years earlier, with a tri-nation race to the pole launched in the first years of Queen Victoria's reign. This book tells the story of these forgotten nineteenth-century expeditions, and of the vast ice kingdom that almost swallowed them whole.

Land of Wondrous Cold also speaks to our present anxiety over global warming and the melting of Antarctica's glacial rim. A long history of Antarctic climate change requires the interweaving of multiple, complementary timeframes. When the Victorian polar explorers converged on the high southern latitudes in 1840–41, they discovered a mountainous land of ice that seemed limitless in space and untouched by time. More than a century later, their scientist descendants, scouring the same Antarctic coasts under the aegis of the international Ocean Drilling Program, finally determined the origin of this continent-sized ice cap in a drastic climate upheaval that helped shape modern Earth millions of years ago. This breakthrough deep-sea research of the late twentieth century uncovered how Antarctica's remarkable refrigeration helped reset the global thermostat, air-conditioning the planet to human specifications.

Fast forward to the present, the world's attention is focused on the potential reversal of this epic Antarctic glaciation. Currently, two hundred feet (sixty meters) of sea-level rise is locked up in Antarctica's towering ice cap. If even a small portion of this glacial imperium were to melt, the great port cities of our modern world would sink beneath the resurgent waves. Already, millions of the world's island and coastal-dwelling populations live under the threat of rising seas. With Antarctica looming once more as a central player in the human story, this book tells the full history of our climatic connection to the white continent—past, present, and future.

In January 2017, I traveled to Antarctica in the mode of the nineteenth-century explorers—by ship. I grew up in the Southern Hemisphere, on that temperate coast where the Antarctic and Australian landmasses were long ago united; but, like most Australians, I had been raised facing resolutely northward. Only after more than a decade as an environmental writer in the United States did my thoughts turn to southerly explorations and polar climes. Encountering, as the Victorian voyagers did, a vast ice pack off the coast of the Antarctic Peninsula, the limitless white plain appeared to me fully alien, a thing that could not possibly intersect with human interests.

For an author, the most forbidding challenge presented by Antarctica's glaciated shore, and the snow-crested ranges beyond it, was its inhuman blankness. Like the Victorians, I was mostly confined to the ship, with nowhere to land along an endless coast buttressed by sheer cliffs of ice. As I gazed out day after day from an icy deck onto a pure, unchanging vista, the metaphor of the blank white page appeared to me really too obvious, but disabling nevertheless. Luckily, on our very last day in Antarctica, on a rocky cape at the northern rim of the Ross Sea, my shipmates and I were treated to a day to write home about.

Cape Adare, at the foot of the Admiralty Mountains, is usually a storm zone charged with dangerous ice. Here Captain James Ross and the crews of the British discovery ships *Erebus* and *Terror*, sailing south from Tasmania in January 1841, were dumbstruck by their first sight of the Antarctic continent. But where they kept a wary distance from the wave-lashed cliffs, the bay was invitingly blue and calm during our visit. On the beach—a rare commodity in Antarctica—the largest Adélie penguin colony in the world awaited us, its photogenic chicks molting in readiness for their swimming debut. Back from the beach stood an explorers' hut from the dawn of the Heroic Age, 1898–99, when a Norwegian named Carsten Borchgrevink established an onshore party on behalf of the British Empire with the goal of being the first to overwinter in Antarctica. These next-generation polar adventurers succeeded, but were miserable and occasionally mutinous. The expedition scientist died and was buried among the penguins.

As if this history were not tantalizing enough for the modern Antarctic tourist, northward across the bay the Admiralty Mountains rose in unspeakable splendor, threaded with glaciers—and we had helicopters to explore them. As we flew along a glacial valley, skimming the river of ice, the physical dimensions of the frozen continent became palpably real to me. If there was this much ice in a single valley—on a continent bigger than the United States—then it might be possible to conceive of Antarctica as home to 70 percent of the world's fresh water, sealed up in these glaciers, and for thousands of miles inland across the endless, dune-like plateaus.

Back on the ship's deck, I watched the helicopter turn toward the southern mountains, expecting at any moment to see the pilot pull up to avoid crashing into the cliff face. But the helicopter disappeared from view long before I could witness the maneuver. In that moment, I felt a fresh affinity with the Victorians, who had likewise been bamboozled by the dry and dust-free air of Terra Australis, which set objects much farther away than they appear. The epic scale of Antarctica—which was the true revelation of being there—became a few degrees less abstract.

For our shipbound party, a day ashore at Cape Adare was a godsend. Penguins, glaciers, and explorer huts offered limitless photo opportunities, while helicopter flights were Instagram gold. To cap the day, a young couple who had met onboard were married on the beach by the ship's doctor, a ceremony witnessed by a penguin legion wearing nature's own tuxedo. Most importantly, from a writer's point of view, Cape Adare completed the all-important guiding metaphor for a book on Antarctica. My weeks spent skirting the ice pack and the blank, icebound cliffs would represent my telling the deep-time history of Antarctic ice, while the action-packed day on the beach at Cape Adare stood for the human-scaled adventure narrative of the first explorers that I interweave here with my glaciological tale.

To its Victorian discoverers, as well as legions of latter-day tourists, Antarctica has appeared pristine and unchanging. But its miles-long glaciers only masquerade as the still point of a turning world. They *flow*, and have ebbed and flowed for millions of years, transforming Earth's climate and biosphere in their transit. Today, parts of Antarctica are the fastest-warming regions on Earth. But if Antarctica is not the immutable ideal it appears to be to the naive tourist, how much has it changed over time? And when?

Since a meteorite brought an abrupt end to the sweltering Cretaceous period with its "great lizards," the interval of greatest sustained warmth in our planet's history was the early Eocene epoch, which began fifty-four million years ago and lasted six million years—so-called Hothouse Earth or, in official terms, the Early Eocene Climatic Optimum. On Hothouse Earth, the Antarctic coast supported diverse forests of warmth-loving palms and flowering evergreens reminiscent of modern-day New Guinea. Farther inland, at higher elevations, stretched a rainforest canopy and sunlit clearings dense with ferns. Though situated near its current latitude, Antarctica managed to sustain its greenhouse botany through fifty days of polar darkness each year. Baking atmospheric carbon levels and mild winter temperatures kept the forest ecosystem running through the unending night. No landscape could be more at odds with Antarctica as it is now—bleak, frozen, and blasted by winds—a continent-sized icehouse.

The Apollo 17 space mission in 1972 was the first to chart a path between Earth and the moon providing a clear view of modern, glaciated Antarctica. One iconic photograph—taken at a distance of eighteen thousand miles from Earth—underscores the dominance of the south polar ice cap in our planet's current geography. Ours is a glacial world, with a narrow tropical band, and higher latitudes and elevations bound by ice (the reason average global temperatures may not suggest icehouse conditions is because we are currently enjoying a brief interglacial period, determined by Earth's orbit of the sun). The Apollo image—one of the most reproduced in history—became known as "Blue Marble." But with signature white awnings spread out at the poles, "Glacial Earth" would be as fitting.

Antarctica's glaciation played a decisive role in the creation of the Blue Marble: our modern, human-habitable planet. Beginning in the 1970s, sedimentary cores drilled in the Southern Ocean floor have revealed the deep climate history of the white continent, in particular its critical shift from hothouse to icehouse. First, in the Mid-Eocene, Australia slipped its hold on Antarctica and drifted northward, creating a cold, globe-girdling ocean in the south. Then, thirty-four million years ago, the end of the Eocene was marked by the most drastic change of climate on Earth since an asteroid suffocated the dinosaurs. Torn from the warming embrace of its ancient Gondwanan neighbors, Antarctica was gradually transformed from

ERA	PERIOD		EPOCH (and Ma)		
CENOZOIC	QUATERNARY		Holocene	0.01	
			Pleistocene		Late
					Middle
				2.54	Early
	TERTIARY	Neogene	Pliocene		Late
				5.33	Early
			Miocene		Late
					Middle
				23.03	Early
		Palaeogene	Oligocene		Late
				33.09	Early
			Eocene		Late
					Middle
				55.8	Early
			Palaeocene		Late
					Middle
				65.5	Early
MESOZOIC	CRETACEOUS		Late	99.0	
			Early	144	
	JURASSIC		Late	159	
			Middle	180	
			Early	206	
	TRIASSIC		Late	227	
			Middle	242	
			Early	248	

Fig. I.2. After the climatic optimum of the Middle Eocene, when Earth experienced hothouse conditions, the world has generally cooled, with a dramatic temperature plunge at the Eocene-Oligocene Transition, about thirty-four million years ago.

a humid land of forests, swamps, and beaches, rich in exotic creatures, into a frozen mountain fortress almost destitute of terrestrial life. Meanwhile, the world's volcanoes abated, lowering CO_2 levels in the atmosphere, cooling the planet a further 5°C and more.

This extraordinary temperature depression—combined with the dynamic power of the new circumpolar Southern Ocean—recast the thermal character of the world. The now worldwide supercooling event devastated the globe's plant and animal life. Everywhere, dense forests gave way to grassland plains, driving an entire weird menagerie of ancient mammals into extinction, from western Europe to the steppes of Asia. Time-lapse photographs from space would have shown an ice cap expanding from isolated flecks of white to the size of a great icy hand holding up the world.

This transition from the Eocene hothouse to the glacial Oligocene epoch was brutally sudden, with over half the temperature change occurring in fifty thousand years. For a planet four and a half billion years old, it was like waking up in sweat-soaked sheets one morning, only to be shivering under the blankets that same night. Earth's shock climate deterioration spelled extinction for a multitude of hothouse creatures and opened the door for a new order of cold-tolerant species: our human precursors.

The overall territory for primates shrank dramatically during the climate crash, but prehuman anthropoids in Africa flourished in habitats abandoned by their thermophilic rivals. A fresh zoology of mammals appeared—a distinctly modern retinue of horses, dogs, and ruminants that our opportunistic ancestors would later domesticate. Human-friendly plants, too—the forerunners of modern cold-resistant grains, and the grass our cattle would eat—emerged from the global cooling event thirty-four million years ago, the age of Antarctica's first ice.

Scientists call it by various unpoetic names: the Eocene-Oligocene Transition; the Oi-1 Glaciation; or, more satisfyingly, La Grand Coupure—the "Big Break." If you were to take a time machine back thirty-five million years, before the Big Break, you would stumble into an unrecognizable faunal extravaganza, with CO_2 levels a sweltering thousand parts per million. Before long, you'd be eaten by a giant prehistoric bird or rat. Now punch in thirty-three million years. As you step out, a strange world greets your eye, but it is not one you mistake for another planet. The atmosphere, most importantly, is accommodating: CO_2 concentrations have dropped from their Eocene highs to modern levels. This is a planet you might, with luck, inhabit. La Grande Coupure was, remotely considered, *our* big break. This book tells the story of the Big Break—of Antarctica's original glaciation, and the planetwide revolution it triggered—through the eyes of a notable band of human beneficiaries: the Victorian south polar explorers and the modern ice scientists who sail in their wake.

To call the Victorians latecomers to Antarctica's story is a spectacular understatement. As a stand-alone continent, Antarctica is thirty-four million years old, whereas our modern ancestors ventured out of Africa a mere sixty thousand years ago. The narrative is familiar. Forests rich in hardwood provided tool-adept humans with the raw material for building ships, which soon crisscrossed the globe: first Polynesians in the Pacific, then Europeans across the Atlantic. Ocean transit opened the world's continents, and even its remote islands, to colonization—with one notable, frozen exception.

Then, at last—less than two centuries ago—the prospect of whale oil, sealskins, and treasures unknown lured ships from Britain, France, and the United States to risk an attempt on the South Pole. The 1838–42 Antarctic discovery missions were the

Apollo moon shot of the nineteenth century—and might never have happened at all. What began as a hopeful idea floated by a few well-connected merchants and scientists snowballed into a full-blown competition with national honor at stake: a race to the South Pole. France sent the brilliant Pacific navigator Dumont D'Urville, while Britain chose its Arctic veteran James Clark Ross to outdo the French. The Americans, meanwhile, with no hall of fame explorer to call upon, gambled on an untested surveyor named Charles Wilkes, whom a certain novelist of renown would later redub "Captain Ahab."

The tri-nation discovery voyages of 1838–42 were the first official expeditions to the Antarctic. But, in another sense, they were also the last of their kind. There is an air of Alice's white rabbit about these Victorian-era explorers, running late for their date with history. By the late 1830s, the four-hundred-year history of European seafaring exploration—begun in the days of Magellan and Columbus—was petering to an end.

The heroics of Scott, Shackleton, and Amundsen in the early 1900s have long overshadowed the remarkable history of Victorian Antarctic discovery. In the much-recycled stories of the Heroic Age, Captain Scott appears larger than life against the backdrop of a brutal polar wilderness, while Shackleton overcame a crushed ship and all that pack ice by sheer force of human will. But Amundsen, to his credit, knew public neglect of the Victorian voyages was a travesty: "few people of the present day," he wrote in 1914, "are capable of rightly appreciating th[ese] heroic deed[s], this brilliant proof of human courage and energy . . . these men sailed right into the heart of the ice pack, which all previous explorers had regarded as certain death. These men were heroes—heroes in the highest sense of the word."

A century after the Edwardian dramas of Scott and company, the Victorians' turn has come around again. The first south polar generation—D'Urville, Wilkes, and Ross—pitifully exposed in their wooden sailing ships and awed into submission by the polar landscape, are explorers custom fit for our current era of climate anxiety. As the Victorians learned, there is no better place to feel the hollowness of fame—as well as time and space—than the looking-glass world of Antarctica. In *Land of Wondrous Cold*, their wandering ships pop up here and there, like Alice's white rabbit, across the vast space-time fabric that was Terra Incognita Australis. Their experience in Antarctica was something closer to what we newly realize today: that human heroics amount to little compared with the greater planetary motions of the continents and climate.

To pursue this point a step further: in my telling of the Victorian discovery voyages, the explorers themselves do not play an outsized role, like actors spotlighted on a stage. Rather, my goal has been to adjust the telescope and bring humans and nature into focus at their proper scale. I recount here the story both of Antarctica's first refrigeration—the origin of its ice sheet—and, millions of years later, the first human encounter with that world-changing phenomenon. *Land of Wondrous Cold* thus interweaves a science-rich story of glaciation and climate change with a more conventional discovery tale set in the South Seas—an unusual blend, admittedly. Any structural resemblance to *Moby-Dick* is strictly intentional.

Antarctica, throughout, is my lead protagonist. The polar explorers of 1838–42—their ambitions, suffering, and wonderstruck observations—are the lens of this Antarctic history, not its subject. Instead of detailing the full story of each expedition (which has been done satisfyingly by others), *Land of Wondrous Cold* re-creates key episodes that link these discovery voyages to the modern era of polar research—to our current understanding of Antarctica's precarious glacial history. Polar science today is booming, courtesy of global warming, glacial melting, and the threat of rising seas. The Victorians' contributions to this new "heroic age" of Antarctic research have long been overlooked, while the harrowing extremes they endured for their exotic polar specimens, observations, and charts make for a discovery legend undeservedly obscured by the expert mythmakers who followed in their wake.

Because Antarctica's discovery was, at its heart, a mythic seagoing adventure—pushing southward beyond the known—my narrative is organized by space rather than time. Beginning with the windswept sub-Antarctic islands, each episode in *Land of Wondrous Cold* takes us degrees farther south toward the elusive pole, deeper down the rabbit hole, curiouser and curiouser. Time, by contrast, is elastic—sometimes spanning eons in a single sentence, sometimes contracting so that whole pages tell the story of a single, crowded hour in Antarctica's 1840s discovery. I invite my readers to speed up or slow down as they wish.

In this telling, the south polar discoverers are not heroic men of destiny. Faced with an alien land utterly inhospitable to humans and resistant to conventions of discovery, the Victorians make no meaningful conquests and plant flags only for show. Instead, they emerge as exemplary slow tourists, absorbing the Antarctic environment at an observational pace unobtainable today. Antarctica, they will discover, tells a story far larger than any single explorer—larger, indeed, than humanity itself.

What has been mostly forgotten, even by polar scientists themselves, is that these discovery vessels of the Victorian Age—the British *Erebus* and *Terror*, the French *Astrolabe* and *Zélée*, and the American flagship *Vincennes*—brought with them the first human beings ever to venture into and beyond the Antarctic ice pack for the purpose of scientific inquiry. What they found baffled, fascinated, and horrified them. They charted ambiguous coastlines, sketched glaciers, collected tiny marine creatures and great seabirds, gathered weather data, monitored the effects of cold on their suffering bodies, and theorized about the currents of the great Southern Ocean. Their collective achievements culminated in arguably the most monumental geographical discovery of the nineteenth century: the Ross Ice Shelf, a great white plateau the size of France, rising sheer from the blue polar waters of West Antarctica from beneath the southernmost active volcano on Earth. In return, the explorers barely escaped with their lives (in most cases).

"Antarctica" is the name of both a continent and an ocean, as well as a more ineffable idea of human limits. It is host to great congregations of seals, birds, and the iconic penguins but has no indigenous peoples. For us humans, Antarctica is not a home but a land of science and imaginative journeying. One hundred eighty years ago, sailing ships from Britain, France, and America burst through the perennial belt surrounding the last undiscovered continent. They entered a hostile glaciated realm whose creation, more than thirty million years ago, helped fashion the planet we inhabit—its climate, ocean currents, and creatures. Now, when Antarctica's melting ice sheet threatens baseline conditions for the human civilization that sent those ships, the land of mist and snow beckons us again. In *Land of Wondrous Cold*, our newly urgent encounter with Antarctica begins with the resolute but forgotten icemen of the Victorian Age.

PART ONE

⚓

Beginnings

⸸ 1 ⸸
The Race Is Joined

On the morning of September 6, 1837, the French Antarctic discovery expedition—consisting of the twin corvettes *Astrolabe* and *Zélée*, manned by a combined crew of one hundred sixty—set sail from beneath the white cliffs of Toulon. The newspapers had declared Dumont D'Urville's expedition doomed, and a small-boat flotilla of the crews' families bobbed alongside the corvettes in sad procession. Mothers and wives wept openly, not without cause as it turned out. By the time D'Urville limped back to Toulon three years later, a quarter of his men had died or deserted. He himself returned a shattered man with not long to live. Seventy years before Captain Scott, Antarctica had martyred its first explorers.

Meanwhile, in Washington DC, French ambitions in the Southern Ocean—and the impressive résumé of Dumont D'Urville—were much on the minds of the Jackson administration. In the 1830s, the American frontier still lay on the oceans, rather than to the West. Andrew Jackson had leafed through the lavish volumes memorializing D'Urville's 1828–29 Pacific discovery voyage in the *Astrolabe* and was enchanted by the exquisite illustrations and general air of heroic enterprise. He declared the United States would launch a discovery expedition on an even greater scale, with scientific wonders like D'Urville's to show for it. That was in 1836. But two years later—with Jackson retired from the White House—the US

Exploring Expedition still languished in port at Hampton Roads, Virginia, the victim of Navy Board infighting and general incompetence.

The French expedition had been announced to the world through the Paris newspapers in June 1837. The following summer, a few days before American commander Charles Wilkes at last gave the order for the US Antarctic squadron to depart, the dismaying news broke that D'Urville had already reached Tierra del Fuego, at the tip of South America, and was headed south. D'Urville had left a note of his intentions in the loneliest postbox in the world, on a hill near Cape Horn. A New England whaleship captain had found it there and dutifully brought it back to Boston. The frustrated Americans—who had been the first to announce their Antarctic ambitions—were now a full year behind the French.

But whatever the Americans' anguish over D'Urville's head start, it could not compare to the hand-wringing in London that summer of 1838—the first of Victoria's reign—where managers of the world's superpower faced the humiliation of conceding to both France and the upstart United States the discovery of the South Pole. British scientists, led by an army engineer named Edward Sabine, had long pushed for an Antarctic mission to complete the charting of Earth's magnetic field. But Sabine had never managed to bring his quixotic plan to the attention of Her Majesty's cabinet, let alone the prime minister's desk.

News of the D'Urville and Wilkes expeditions changed all that. By November 1838, a South Pole expedition was the subject of urgent discussion at the palace in the presence of the young queen and her first of many prime ministers, Lord Melbourne. A meeting with the chancellor of the exchequer followed. With almost unseemly speed, a hundred thousand pounds materialized to finance the royal expedition in quest of

Terra Incognita Australis. Many thought it beneath the British, as the world's supreme naval power, to compete with French and American ships in any oceangoing enterprise. But here they were, drawn into a race to the South Pole. Once committed, there were no half measures. The Admiralty sent their best vessels—the ice-breaking ships *Erebus* and *Terror*—and a bona fide polar hero, James Clark Ross, celebrated discoverer of the North Magnetic Pole, to command them.

But even the rulers of the British Empire could not make time slow down. A distant third down the rabbit hole of the Antarctic chase, Ross reached the subtropical island of Madeira off the coast of Africa in October 1839. His officers climbed its celebrated peak for a view of the Atlantic Ocean and looked about for a small pyramid of stones—a message cairn—left by their rival Wilkes. The American's note had been eaten by goats, but the *Erebus* officers learned from the locals of Wilkes's plans to sail south that summer. With Antarctic sea ice in retreat only three months a year—December through February—the South Pole was beyond the reach of the British that season. Through no fault of his own, Ross now found himself a year behind the Americans and two behind the French. The cloudy prospect of defeat hung over him and his expedition. Reviewing his options at Madeira, Ross decided the farthest south he could venture that year would be the sub-Antarctic waters of the Indian Ocean—to Kerguelen Island, perhaps the remotest anchorage on Earth. From there he would make the long odyssey east to the Australian port of Hobart, to gain intelligence of French and American movements and scheme to outdo them by whatever means necessary.

Back in the spring of 1837, King Louis-Philippe had surprised Dumont D'Urville by approving his request for a third discovery expedition to the South Seas. But the famous explorer's delight

Fig. 1.1. Posthumous portrait of Dumont D'Urville, by Jerome Cortellier (1846).—Grand Palais, Chateau de Versailles.

was short-lived. His Majesty stipulated that D'Urville include, in his three-year cruise, an attempt on the South Pole to plant the French flag ahead of the British and Americans, and chart new killing fields for struggling French whalers. D'Urville was dismayed. He admired the British polar explorers—Cook, Parry, and John and James Ross—but he would take three years in the tropics over two months in the ice.

No one knew what lay beyond 74° south: an open sea, a continent filled with unheard-of creatures, perhaps a giant abyss? D'Urville, a man of science and adventure his entire adult life, had never seen this particular mystery as his to solve. He deferred to his hero, James Cook, who had halted before the southern ice fields sixty years before and declared it a place of horror no man would wish to penetrate.

Why the South Pole? Why now? The king, he guessed, had been reading the adventure narratives of two celebrity whalers: Englishman James Weddell and American Benjamin Morrell. Both authors offered tantalizing prospects of an ocean gateway to the antipodean pole, tropical temperatures, and virgin land beyond the moat of ice that had deterred Cook, with plentiful stocks of whales and seals.

In preparation for the voyage, D'Urville visited London to learn what he could and to acquire the latest charts and compasses from the capital of the seafaring world. The Admiralty officials he met were tight-lipped about the far south, resentful that a French expedition should presume to explore in "British" waters. On the subject of James Weddell's voyage of 1823, however, they were suddenly eloquent. Captain Weddell was a "true gentleman," his record-setting push to 74° south a triumph of British seamanship. The French public had devoured Weddell's story in translation. But D'Urville, who had never heard of a gentleman sealer, was unconvinced. And as for "The Greatest Liar in the Pacific"—Benjamin Morrell—the nickname had been well bestowed. But no one at the French court had had the wit to keep these adventurers' tales out of the hands of the king, or courage enough to tell him it was all fantasy.

On his previous two journeys around the world, D'Urville had been robust and full of hope, the emperor of his own illusions. But now he was an old man (in sea years), chronically ill, and disenchanted with the world. The long summer before the Antarctic expedition sailed, he felt himself losing sight of his destiny. In his dark moments, he felt he was leaving his beloved wife and sons for a years-long mission to the ends of Earth for an impossible reward.

D'Urville had courted Adéle Pépin in storybook style, lingering in her father's Toulon shop while she smiled at him from behind the counter. But in their Toulon villa that summer, the veteran explorer and his wife were like a duet of doleful violins, played in separate rooms. D'Urville had used Adélie's weakness for explorer romance against her, and touched her maternal ambition. With this third voyage, he told her, he would inherit the mantle of the great Cook. Their children's futures would be secured as sons of a national hero. She had given in to his will—as they both knew she would—but at a cost. As the day of his departure approached, she withdrew behind her habitual veil of sorrow for the two children they had lost. Husband and wife spoke less, and although Adélie did not complain, he knew she felt that he had trapped her, that he was sealing for her a disastrous fate.

D'Urville's anxiety at the prospect of an Antarctic voyage became the stuff of nightmares. The dream would begin with Cook and himself side by side before some great parliament, basking in the fame of their three voyages around the world. Then he would be back in the ice at the helm of the *Astrolabe*, in a narrow strait with dead ends fore and aft. He would shout hoarse orders to an empty deck, while the ice renewed itself perpetually ahead of him—a deadly passage with no way out. But when his official orders to sail to the pole arrived from Paris, the nightmares abruptly ceased. From the moment the old campaigner set foot on the deck of the *Astrolabe*, all disabling domestic feelings vanished. He became his orders. Adélie, and his sons Jules and Emile, bid him a tearful farewell on the dock. The French polar expedition launched just in time. Two weeks after the *Astrolabe* and *Zélée* set sail from the harbor, a deadly cholera outbreak struck Toulon.

How did D'Urville reconcile himself to this Antarctic mission? It being 1837, with the great age of exploration in the past and vulgar commercial interests now ascendant, the fickle French public demanded novelty. "The people like surprises," Napoleon had said. A journey to the wild unknown of Antarctica was definitely that. But there was a nasty undertow to contend with, namely, *resentment in high places*. His longtime enemy, François Arago—director of the Royal Observatory and member of the Chamber of Deputies—had denounced the polar expedition as an expensive folly. D'Urville was leading his men to certain death in the ice. "When we have spent the people's money this year to send him to the desolate ends of the Earth where there is nothing to discover," Arago raged before the chamber, "will we then have to vote next year for funds to fetch the bodies back?"

With Arago's prophecy of doom published in every newspaper in France, D'Urville's recruitment drive on the Toulon docks faced a sudden crisis. He was forced to fill his complement for the *Astrolabe* and *Zélée* with novices, boys, and hangers-on. Some were too stupid or dangerous-looking to accept, even for a man-hungry captain. And most that did pass muster were rated third class—unskilled and untested—for a voyage demanding the best men. Such was the shabby reality of exploration, not mentioned in the wonderful travel narratives his wife Adélie devoured (along with half the rest of Europe). To solve his manpower problem, D'Urville petitioned the king to offer prize money: one hundred francs a man to beat Weddell's record southing, and ten more for every band of latitude beyond it. This was a patriotic appeal the Toulonese seafaring community could not resist.

Meanwhile, across the Atlantic, the US Navy could not boast of a Dumont D'Urville or James Clark Ross. In fact, the secretary of war, Joel Poinsett (of "poinsettia" fame), found himself unable to persuade *anyone* to take command of the US Exploring Expedition. His predecessor Mahlon Dickerson, an Olympian footdragger even by Washington standards, had undermined the best candidate, Thomas ap Catesby Jones, a stalemate that ended only with their twin resignations.

Captain after captain turned Poinsett down, spooked by the Jones affair and continuing bad press. A Navy mission that had begun with patriotic promise was now beset by corruption and uncertainty. The newspapers labeled it the "Deplorable Expedition," as recriminations in Washington mounted. At last, a little-known lieutenant surveyor named Charles Wilkes had greatness thrust upon him. He hurried home to his wife Jane with the news, where they cried for joy at his snatching the prize. In light of what followed, they might have saved a few tears for the aftermath. Wilkes's worst enemy could not have contrived a means of ascent better guaranteed to undo him in the end.

Having lost his mother early, Wilkes spent his childhood neglected by a succession of nurses and schoolmasters, who taught him useful lessons in brutality. For a while, he was boarded at the school of a Mr. Smith, in what was then the town of Greenwich, north of Manhattan. Wilkes, the son of a gentleman, belonged to a gang of upper-class boys in the neighborhood—the sons of the Livingstons and De Puysters—who lacked the courage to assert their territorial rights over the butchers' boys and their like. The younger boys, led by Wilkes, grew tired of the harassment and were ashamed for their unwarlike older brothers. At last, one day, a pitched battle was launched with bricks, bottles, and knives. The conflict stood in the balance until little Charley Wilkes hurled a missile at the opposition's Goliath, an illiterate errand boy named Moore, and felled him in the mud. While Moore's life hung

Fig. 1.2. Portrait of Charles Wilkes, by Thomas Sully (1840).—US Naval Academy Museum.

in the balance, the authorities investigated. No boy would name the assailant, until Wilkes himself stepped forward. To his dismay, no one believed him on account of his small stature. He was prepared for any punishment so long as his heroism was recognized. But it was not. The errand boy recovered; Mr. Smith's school shut down; and young Charley moved on yet again. Decades later, the incident still rankled Wilkes, who would battle until the end for the fame he was owed.

On the eve of the American squadron setting sail, Lieutenant Wilkes had bought an expensive new coat, emblazoned with captain's epaulettes and brass buttons, only to learn he was not to be promoted. To fold the coat away in his chest was agonizing. It was a critical moment in the history of the US Exploring Expedition. By refusing him the authority of a captain's rank, the Navy fatally undermined their man. The denial of promotion fed Wilkes's own sneaking fear that he should not, in fact, be leading this fleet of ships around the globe, to the South Pole and back. A tense and angular man with no talent for command, Wilkes would soon be overwhelmed. Already a small crack opened in his brain for panic to seep in.

Wilkes could take no comfort from his officers, or from the scientific gentlemen on board, who seemed to view what lay ahead as a round-the-world pleasure cruise. They anticipated exotic sights, meeting natives, and collecting rocks and butterflies. They furnished their cabins with carpets, curtains, and chintz in Babylonian abundance, alongside tea services and bookshelves. These were the accommodations of gentlemen embarking on the Grand Tour, not sailors headed to the Roaring Forties and beyond, where waves built up like mountain ranges for thousands of miles, and every squall was a potential killer. Another crack opened in Wilkes's mind, this time for bubbling rage.

Then news filtered down from Boston of the Frenchman D'Urville, who had left letters in Tierra del Fuego announcing his imminent departure in quest of the South Pole that February. The Americans had been the first to announce their plans for Antarctic discovery but had squandered their advantage. Now the Frenchman—experienced and determined—had a year's head start. A fresh crack opened—more nervous agony for Charles Wilkes.

When, on the morning of August 18, 1838, the inexperienced American commander at last gave the order to make sail out of Hampton Roads on the Virginia shore, excitement rippled through the squadron. A decade of bad feelings was temporarily forgotten, and ambitions for the Exploring Expedition—called the US Ex. Ex. for short—stretched to the very horizon. The officers on board the flagship *Vincennes* wore broad smiles as they waved to the crowds of well-wishers lining the banks. After all, they belonged to the largest (and *last*) voyage of exploration under sail ever launched from the Western world. Of the 350 men of the US Ex. Ex. setting forth on their epic journey that bright August day, only Charles Wilkes believed them destined for disaster.

Unlike Dumont D'Urville with his light-sailing corvettes, and Charles Wilkes with his leaky flotilla, James Ross had been handed the two best ships in the world for polar exploration. *Erebus* rode the waves like a duck. Or when the sea did mount the decks, she staggered only a little under the weight of water before a porthole knocked out to release the torrent and she steadied herself again. Admittedly, her companion ship *Terror* had been beset for ten months in Hudson Bay and recently wrecked, but repairs were now complete.

Erebus and *Terror* had been designed for warfare, specifically to fire three-ton mortars at Napoleon's navy. To survive the recoil of giant guns, each ship was triple-hulled oak, several feet thick at the waterline. Her copper sheathing was doubled around and four sheets thick at the bow. To safeguard against an accident with their bombs, both *Erebus* and *Terror* had their holds divided into three separate, watertight compartments. These were further subdivided by thick double walls of oak and teak. Even the decks were double planked. During the War of 1812, *Terror* participated in the bombing of Fort McHenry, but a decades-long peace had enabled *Terror* and *Erebus* to be repurposed as icebreakers, Victorian style.

The British Admiralty, accustomed to sending expensive expeditions to the Arctic, had not been stingy with the South Pole mission. In the summer of 1839, Ross left the business of fitting out and provisions to his Arctic cabinmate and trusted second, Francis Crozier, while he stayed in London tinkering with magnetic instruments under the nervous eye of Edward Sabine. It had been Crozier's thankless job to bribe and cajole the little emperors of the Royal Navy dockyard at Chatham for the proper quantities of cordage, canvas, et cetera. At Ross's instructions, he had also issued bulk orders for food stored in tin cans, a new technology. Between them, the ships' holds now contained thirteen and a half thousand pounds of meat, topped by five thousand pounds of gravy, along with fifteen thousand pounds of tinned vegetables and six thousand pounds of soup—enough food for four years at sea. Having recently spent that many years marooned in the Arctic Circle himself and made a close study of the native diet, Ross set great value on fatty foods to insulate a sailor's body against the cold.

A decade before his maiden voyage to Antarctica, Ross was already the most experienced polar explorer alive and the owner of a celebrated scientific discovery in the Arctic. Then in his mid-twenties, he had spent a third of his life on the northern ice. He had learned the language of the Inuit people, and how to manage a sled with dogs. When the long winter darkness lifted at last, in the middle of May 1831, he led a sledding party to plant the British flag at the North Magnetic Pole.

His small band came to a stretch of low ground near the water. They had reached 82° north, at a point farther west in the Arctic than any previous expedition. Inland, a mile distant, the beach rose to a low line of sandy ridges. Here Ross took out his dipping needle, careful with his numb hands. When released from its horizontal position, the needle swung with total confidence in a 90° arc, pointing down to his feet. The built-up anxiety of the two-year Arctic voyage—and five days of raw suffering on the march—melted away. The fact that he and his men had not eaten for twenty-four hours—with their food another long day's trek away—was forgotten. Standing at the North Magnetic Pole, James Ross felt a transcendent happiness.

Some had expected the party to find a great mountain of iron at this geographically momentous spot, or a giant magnet the size of Mont Blanc. But Nature had chosen not to advertise the locus of her "great and dark powers." It was as bleak a place as could be imagined, without food, fresh water, or vegetation. The rude Earth. Notwithstanding, they raised the flag at the magnetic pole and gave three cheers for King George, not realizing he was dead. In that moment, James Ross's geographical destiny—to become the first man to stand at both ends of Earth—was marked. As the Royal Society put it, "It is not to be supposed that Captain Ross, having already signalised himself by attaining the northern magnetic pole, should require any exhortation to induce him . . . to reach the southern."

But would James Ross survive to make that south polar odyssey? For two long years, it seemed he and his shipmates would not. When his uncle John Ross—the commander of their ice-locked vessel—sat down in his cabin to compose his annual report to the Admiralty on New Year's Day, 1832, he admitted his letter had little chance of reaching its destination: "I confess that the chances are now much against our being ever heard of."

Everyone else was asleep when, on a freezing summer's morning in 1833, James Ross spotted a sail on the horizon. For a few heart-stopping hours, it seemed the ship was preoccupied with whales, but at last she saw them. When the boat pulled alongside, the whaler's mate gaped at the spectacle of two

Fig. 1.3. Portrait of James Clark Ross, by John Wildman (1834). Ross is pictured, in full heroic regalia, on his return from the Arctic as the celebrated discoverer of the North Magnetic Pole. He was too poor, however, to accept the king's offer of a knighthood.—National Maritime Museum, Greenwich.

dozen skeletal men—bearded, gaunt, their faces blasted. Some were lame, one blind. Their rescuer told them his ship was "the *Isabella* of Hull once commanded by Captain Ross." Uncle John Ross had been expecting death. His old war wounds were open and bleeding. But here was his old ship come to save him. "I'm Captain Ross," he replied hoarsely, only for the young mate to explain that he couldn't be as the old man had been dead these past two years at least.

On his return to England, James Ross had done his best to sabotage his friend Sabine's plans for a South Pole voyage by falling in love. Her name was Anne Coulman—a girl of seventeen—the daughter of a wealthy friend of his sister living in Yorkshire. Ross, now in his thirties, was instantly smitten.

To Anne, her explorer-lover must have seemed impossibly glamorous. As for their age difference, it seemed not to enter her thinking. But her father was another matter. To Thomas Coulman of Whitgift Hall in Yorkshire, Ross's fame and good looks meant less than nothing. His daughter's suitor was a penniless adventurer. With no war on the horizon, his opportunities for advancement were near nil. A man could take himself to the ice and maroon himself for years, but it did not mean he had money to marry. Rumor had it that Ross had refused a knighthood from the king because he could not afford the fees for regalia. And if he never returned from this latest escapade to the South Pole, as was probable, he would leave his Anne a broken-hearted widow. The fact that Ross had dared to romance a mere schoolgirl right under his nose especially rankled. Coulman banned the lovestruck captain from Whitgift Hall and kept a close watch on his daughter.

But Ross had a willing go-between in Anne's cousin Jane. Through her, he passed letters and a plan to breach Whitgift Hall by stealth. A convenient stream allowed him access by boat where he waited for Anne's signal from the upper window. A little copse by the water offered the perfect lovers' haven in the darkness. There they made plans together. Although it meant a long separation, Ross must make his exploration to the South Pole. He would contract with Murray, the London publisher, for a large sum to sell his narrative of the voyage on his return. With that money, and lecture fees, they could begin their life together.

By the time the prime minister approved the British Antarctic voyage in 1838, James and Anne had already been in love for more than four years. Their courtship, thus far, had consisted of clandestine letters and a handful of breathless meetings. Now he was to be another four years gone, while Anne waited for him at Whitgift Hall, like Penelope, except that she must pretend to have forgotten her Odysseus. During his tour of the British Isles for Edward Sabine, taking magnetic observations, Ross had circled back to Yorkshire as often as he dared. He kept their engagement a secret from the Admiralty, just in case.

Ross had never before left a sweetheart behind to journey to the ice. When, on his four previous polar voyages, he watched shipmates pine away, he had congratulated himself on his strength of character and good luck. But now he was being paid in kind. During the *Erebus*'s long haul south in the closing months of 1839, he sequestered himself in his cabin and wrote long letters of tender encouragement to Anne, whom he had left at the mercy of her resentful father. His officers felt the neglect and began to question their captain's gold-plate reputation.

Everyone in the Royal Navy knew the stories of Captain Cook's bouts of gloomy madness in the South Seas (Cook's martyrdom on the beach at Hawaii ensured those stories were not made public). When Ross felt his mind giving in the same direction, he took positive steps before it was too late. During his Arctic voyages with Edward Parry, his captain had encouraged young Ross's interest in science, though he had little formal education. In addition to magnetism—which made use of his flair for mathematics—he had pursued the study of marine animals. He became adept with a tow net and bottling specimens in spirits. He had even ventured to publish some brief accounts of his findings on their return.

Now, tormented with longing for Anne and for home, the daily discipline of natural science came to his rescue. Whenever the *Erebus* was at anchor, Ross took out the small boat and, armed with a bucket, waded barefoot in the shallows and collected specimens for hours on the beach. At sea, he ordered tow nets to be cast off the stern and dredged the ocean floor with the sounding line for whatever corals or fossils he could find in the freezing muck. By very good luck, he discovered a highly capable colleague in his young assistant surgeon, Joseph Hooker. The mechanical labor of collection, followed by the daily reward of contemplating the line of spirit jars that stretched farther on the shelf of his cabin, meant that the long days on ship, in their unvarying procession, passed less painfully than before. In the patient handling of delicate sponges, mollusks, and tiny squirming crustaceans—bucketful after bucketful—Ross found the outlet for a romantic despair he thought would kill him.

Interlude: The Hollow Earth

As 1838 dawned, no one knew what the three exploring expeditions would find beyond the Antarctic Circle. But one bold speculation on polar geology would prove remarkably persistent, even after their discoveries, and still may be found in rogue corners of the Internet today: the Hollow Earth theory.

Hollow Earth was a touchstone of Victorian popular imagination. In the humorous poem that opens *Alice's Adventures in Wonderland,* Lewis Carroll remembers taking his colleague's daughter Alice and her two sisters out in a rowboat on a summer's day in Oxford in 1861. As usual, the girls demanded that their father's amusing friend tell them a story "with nonsense in it." For Carroll (whose real name was Dodgson), satisfying the Liddell sisters' appetite for stories was both a pleasure and a burden. In *Alice in Wonderland,* he imagines himself as the narcoleptic dormouse at the Mad Hatter's tea party: he would sometimes fake sleepiness, like the dormouse, to avoid having to entertain the relentless threesome anymore.

But in the rowboat that day, there was no escape. So Carroll ransacked his memory for snatches of fun to cobble together a story. Fortunately, a lifetime's consumption of word games, math puzzles, and pantomime shows was at his disposal. From this rich mental archive of Victorian popular culture, an oddity from his childhood popped into his head: back in the 1830s, a faddish idea circulated that Earth was hollow, with openings at the North and South Poles and exotic peoples living inside. What if he shifted the northern entrance of this hollow world to a place the girls knew—say, in the nearby woods at Gosford—and told a story about Alice falling down a rabbit hole through the center of the planet? Alice could talk to herself as she fell, and prepare to greet the strange people of the Antipodes—Alice would call it "The Antipathies"—in New Zealand, Australia, and beyond. What adventures Alice could have among the "upside down" people of an alternative southern hemisphere! From that moment's desperation in the rowboat, the White Rabbit, Cheshire Cat, and Tweedledum and Tweedledee were born.

Lewis Carroll's inspired notion that day was to make little Alice Liddell into a full-fledged polar explorer, and send her through a tubular Earth to "Wonderland," the *terra incognita* of the South. Alice's first wish, as she is falling down the rabbit hole, is to become a telescope—an explorer's signature accessory. Later, Carroll will have her neck stretch until she looks just like one. Alice's elongated neck is the storyteller's metaphor for time and space in Wonderland/Antarctica: a rubbery, telescopic world of perpetual mirages, where Alice is alternately miniature and giant, the White Rabbit is always running late, and the native fauna are truly "frabjous," like cartoon relics of the Mesozoic.

Popular science fiction, like *Alice in Wonderland*, originates with the Hollow Earth theory, which blossomed in America with the Jackson-era mania for Antarctic exploration. A war veteran from Ohio named John Symmes launched a national campaign to explore "the holes at the poles" with a pamphlet published in Saint Louis in 1818, which he circulated to newspapers and legislators across the nation:

> I declare that the earth is hollow and habitable within; containing a number of solid concentric spheres, one within the other, and that it is open at the poles. . . . I pledge my life in support of this truth, and am ready to explore the hollow Earth, if the world will support and aid me in my undertaking. . . . I engage we [will] find a warm and rich land, stocked with thrifty vegetables and animals, if not men.

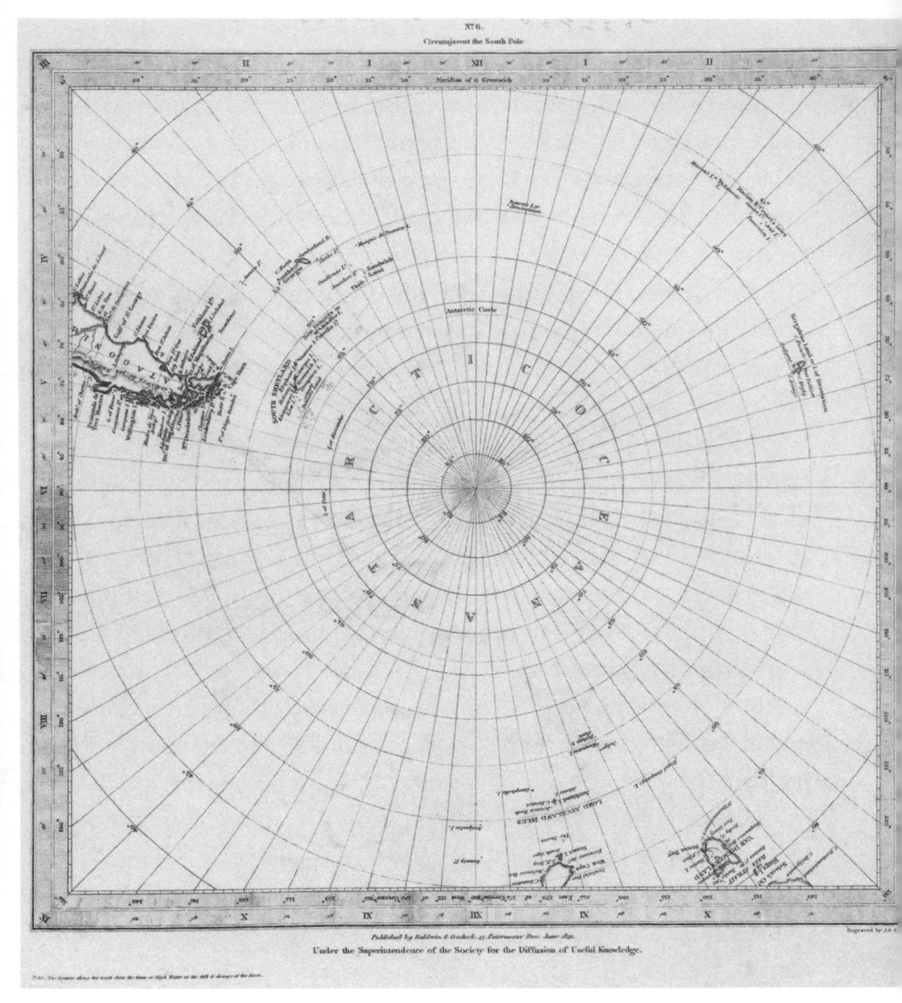

Fig. 1.4. An 1831 map of Terra Incognita Australis, drawn for the Society for the Diffusion of Useful Knowledge, shows what little was known of the polar region prior to the discovery voyages of 1838–42. The conspicuous void gave rise to numerous geographical speculations, including Hollow Earth theory.—University Library, University of Illinois at Urbana-Champaign.

Armed with a wooden model of a hollow Earth, Symmes took to the lecture circuit where, in town after town, he earned a loyal, admiring audience, including the educated.

One convert, James McBride, took it upon himself to legitimize Symmes's cause through publication of a serious-minded Hollow Earth manifesto. Published in 1826 in Cincinnati, McBride's book—titled *Symmes's Theory of Concentric Spheres; Demonstrating that the Earth Is Hollow, Habitable Within, and Widely Open about the Poles*—is a classic of American pseudoscientific literature, bursting with data, excited syllogisms, and epigraphs from Shakespeare and Milton.

After seven chapters describing the physics and geology of Hollow Earth, McBride makes the key connection between Symmes's theory and the necessity for a national polar expedition to verify its truth: "for, should it hereafter be found correct," writes McBride, "the habitable superfices of our sphere would not only be nearly doubled; but the different spheres of which our earth is probably constituted, might increase the habitable surface ten-fold." For a patriotic American readership, the Hollow Earth theory offered fetching visions of colonial expansion and a planet-sized virgin interior awaiting its destined embassy from the New World American metropole.

The Arctic had been crawling with British and Russian heroes of late—reducing the potential dividend of an American mission to the north—so McBride turned his attention toward the unexplored Antarctic region: "the most practicable, the most expeditious, and the best mode of exploring the interior regions would be by sea, and by way of the south polar opening, crossing the verge at the low side, in the Indian ocean, where it is presumed the sea is always open, and nearly free from ice."

Enter Jeremiah Reynolds, a charismatic, would-be polar adventurer who provided a necessary link between the backwoods show circuit and East Coast officialdom. A gifted promoter,

Reynolds first joined Symmes on his tour of the byways of Ohio and Pennsylvania, where he served as frontman for the Hollow Earth cavalcade. To spellbound audiences, Reynolds proclaimed the Symmes gospel that Earth contained huge openings at both poles—was, in fact, a kind of cylinder—and that an American exploring fleet should be sent to claim this second New World.

When "Symmes's Hole" became a national joke, Reynolds dropped him, but not his monomania for a South Pole expedition. He lobbied senators and Navy bureaucrats with relentless energy. In a remarkable coup, he addressed a full house of Congress with a fervent pitch for America's destiny to rival, through oceangoing exploration, the great scientific empires of Europe. And because Reynolds's friends in the frontier west had seen in his efforts a test of their own power, they rallied the House around a bill to authorize the US Exploring Expedition, passed on May 9, 1836.

Among Reynolds's admirers was Edgar Allan Poe, whose susceptible brain was touched with the fantastical possibilities of a vacant planetary interior. Poe, like all Hollow Earthers of his day, desperately wanted Symmes to be right. Even when Reynolds abandoned Symmes, Poe took up the polar cause of his new hero in the Baltimore newspapers. In his debut novel, published as Wilkes embarked, Poe sends his eponymous hero, Arthur Pym, on the same southward course as the American explorers. Murder, kidnap, mutiny, and shipwreck ensue—a Poe-like litany of horror on the high seas.

But the true mark of Poe's Hollow Earthism lies in his 1838 novel's bizarre denouement. His hero Pym's ship sails southward beyond James Weddell's record southing and the icepack, and into an open sea. There, the accidental explorers come upon a new Earth bejeweled with tropical lands, then to the verge of the Symmesian polar opening where the concavity of the planet beckons. For Poe, geographical discovery in Antarctica was never the principal lure. Rather, the thrill of extreme polar exploration lay in the opportunity to be sucked from the surface of the ocean, via a hectic maelstrom, into the existential gloom of a tubular planet. In the last lines of the novel, Pym's ship tumbles into a giant vortex via "a limitless cataract, rolling silently into the sea from some immense and far-distant rampart in the heaven . . . where a chasm threw itself open to receive us." With the hero's ecstatic fall into a parallel inner world, science fiction, the signature popular genre of modernity, is born—via the tropic canal of Hollow Earth.

The postscript to Poe's *Narrative of Arthur Gordon Pym* tantalizingly refers to lost chapters "relative to the Pole itself," that "may shortly be verified or contradicted by means of the governmental expedition now preparing for the Southern Ocean." The very month of the novel's publication in 1838, as Wilkes's squadron prepared to sail south, the poverty-stricken writer wrote to the secretary of the navy pleading for "the most unimportant clerkship in your gift—anything by sea or land." Edgar Allan Poe in Antarctica? But it was not to be—Poe's fantasy of cruising the hollow verge at the South Pole remained in the literary domain, in a vanguard text of the new science fiction that Jules Verne and his successors would later bring to an insatiable worldwide audience.

‡ 2 ‡

Joseph Hooker Travels through Time

By October 1839, the French and American expeditions had already made their first forays to the south polar ice, while their British rivals, under James Ross, still languished north of the equator. At latitude 32° 39′ north, the island of Madeira lies in the main shipping highway connecting Europe to the coasts of Africa and South America. Offering both subtropical warmth and cooling mountain breezes, the island enjoyed perhaps the most celebrated climate in the nineteenth-century world. Legions of invalids from the British Isles—land of catarrh and consumption—sailed to Madeira on doctors' orders to restore their wasted frames. Most ended up buried in the English cemetery in Funchal, a romantic, tree-lined resting place where Joseph Hooker, the young botanist of the *Erebus*, and the sole member of the British Antarctic Expedition to win enduring scientific renown, duly paid his respects.

The beautiful weather on Madeira enraptured Hooker, who had grown up in Glasgow's dank fogs. He walked the decks of the *Erebus* all night, luxuriating in the soft air. For a snack, he spread banana on bread like butter and ate grapes. He inhaled the fragrance of orange groves wafting from the shore. When day came, the cloudless sky—like nothing in Scotland—reflected on a laughing blue sea. In a letter to his sister Maria,

Fig. 2.1. Portrait of Joseph Dalton Hooker, by George Richmond (1855). Hooker is represented as a confident, prosperous young gentleman. In truth, the Ross expedition offered a rare opportunity for the professional distinction in science he desperately needed.—Frontispiece, Leonard Huxley, *The Life and Letters of Sir Joseph Dalton Hooker* (1918). Library, University of Illinois, Urbana-Champaign.

he described it as the exact color of the lapis lazuli ring he had given her as a parting present.

Ironic, then, that Joseph Hooker fell ill with a fever at Madeira, where so many of his compatriots sought good health. His mistake was to climb too energetically from the heat of the town into the crisp mountain air and then to fall asleep in the damp grass. Disturbed by the sudden shift from hot to cold—and who knows what vagrant microbe—his body shook uncontrollably for weeks. In his letters home, he kept his illness a secret from his father, the director of the Royal Botanical Gardens at Kew, who he knew would blame the illness on his son's poor judgment.

If Madeira was paradise, then the Cape Verde islands, at 14° north, were a dusty, equatorial desert, with barely a tree for relief against the penetrating sun. South of Cape Verde, the *Erebus* and *Terror* encountered the dreaded "Variables," the unstable climatic zone between the northeast and southwest trade winds, where torrential squalls alternated with baffling calm, and the explorers sweltered in the heat. The men were shockingly sunburnt, made worse by a prickly red rash. The *Erebus* had not been designed for open hatchways, making conditions below unbearable. When they could, the men and officers slept on the open deck. The cocks and hens they had brought to populate remote islands grew stupid in the heat, forgot to find shade, and died.

Still feverish, and tormented by the hot pimples erupting all over his body, Joseph Hooker grew concerned for his botanical collections. The dampness aboard was universal. For botanical requisition, the British government had supplied two dozens reams of paper for sketching and preserving specimens. Also, two collection boxes—long metal cylinders with a vertical door and shoulder strap—for botanizing in the field. Most important of all, Hooker had two large Ward's cases for bringing plants home alive, through temperatures at all latitudes. But already, several dozen of his specimens had fermented inside their rotting paper presses.

Hooker established a routine of drying his plants in his cabin, and stringing his papers to dry on a line on deck. But a three-day storm out of Cape Verde undid all that work. With hatches battened against the sheeting rain, the moisture below deck grew so thick it ran in rivulets down his face. Compounding these daily anxieties of expeditionary botany, Hooker felt a weighty professional burden on his twenty-four-year-old shoulders: above all, to do better than the scientists the rival French and American expeditions had brought with them. Moreover, given the unprecedented route this Antarctic expedition was to take, no English botanist would ever likely follow in his footsteps. Worst of all, he labored under an all-consuming need to please his intimidating father, the director. If his collections were not worthy of the Hooker name, his life would not be worth living.

On March 7, nearing the African cape—as far south of the equator as Madeira lay to the north—the air turned suddenly colder, and the ships sailed into a Scottish-style mist. By mid-April, the daily sample of seawater, drawn up in an ingenious cylinder, showed a marked decline in ocean heat. Captain Ross speculated on the vicinity of icebergs, and within a week the temperature of both air and sea had dropped thirty degrees. The men, who had itched themselves raw in the heat of the Variables, were loath to give up the convenience of bare feet and cotton and required a positive order from their captain to break out their cold climate stores. Streaming colds and body aches drove many of them to the sick bay.

South of the African Cape of Good Hope, they sailed into the Roaring Forties, the wildest seas on Earth—continual storms, giant swells, and thousands of ocean miles with barely a crop of rock to be seen. When at last they came to Possession Island, in the sub-Antarctic Crozet archipelago, the orange groves of Madeira might as well have been from another world. The *Erebus* dared not approach too near to Possession's rocky coast, where she was due to rescue a party of marooned sealers. This gave Hooker ample time to contemplate the original Antarctic fauna: thousands of penguins standing sentinel on the rocks, bowing and strutting, unconcerned by the freezing spray of the waves. They looked like miniature soldiers in two-tone pants.

The sealers, when they came aboard, seemed to have been penguin-ized. Hooker, a fastidious young man, was shocked at their appearance. No one would mistake the *Erebus* for a perfumerie, but the castaway sealers disgraced the captain's cabin with their filth and stench. They wore boots made of penguin skin and feathers, and their faces were caked in that animal's grease. Since their shipwreck, they had survived on albatross eggs and the flesh of sea elephants, particularly tongue and flippers. Beyond this, they could barely describe their island home or its resources. To Hooker, they seemed barely human.

Looking out from the *Erebus*'s deck toward the barren rock of Possession Island, huddled in his lambswool jacket, the young naturalist began to wonder whether he had made a terrible career mistake in joining the Antarctic expedition. How could he botanize in a climate too brutal for vegetation, in a sea without land? Disappointed in his attempt to make landing in the Crozets, Captain Ross now set sail southeast for the next speck of land in the watery vastness of the Indian Ocean: Kerguelen Island, which Captain Cook had called "Desolation." Hooker's heart sank at the thought.

Back in the last age—in January 1772—King Louis XV had sent an ambitious young naval officer named Yves-Joseph de Kerguelen to the South Seas, to put to rest the endless speculation among his savants of "Une Terre Australe," a Great Southern Land. A Europe-sized continent at the South Pole was scientifically necessary, they argued, to balance out the landmasses of the crowded Northern Hemisphere. Louis had another motivation. With his English rivals gobbling up Pacific Ocean colonies at an insatiable rate, the king felt an urgent need to establish a French satellite dominion in the South.

What happened next defies all reason. Captain Kerguelen, fatally confusing the king's wish with his command, sailed into the cold, high latitudes of the Indian Ocean determined to make his name as the French Columbus. On February 13, 1772, the ship's lookout reported land—high mountains shrouded in fog. For Kerguelen, this was enough. Without waiting even for his sister ship—which actually made landfall—he raced back to Versailles to announce his discovery of "La France Australe," a lush southern continent (he said) blessed with forests, lakes, and a civilized native population devoted to the arts.

When, after a follow-up voyage, Kerguelen was found to have peddled lies of truly continental proportions, the king threw him in jail. The Englishman James Cook, commander of His Majesty's ship *Resolution*, on the same search for a southern continent, landed at "La France Australe" two years later, and found a small, iron-bound island bereft of trees and animals. Cook's "Desolation Island" was the name American sailors knew it by throughout the nineteenth century, where they slaughtered every last seal for their skins to sell on the Chinese market.

History has quietly overlooked this French embarrassment in the sub-Antarctic. More than two centuries later, the disgraced Kerguelen's name remains attached to the erstwhile "France Australe." From one perspective, however, the Frenchman has been vindicated for his bizarre, career-ending fantasy of a verdant southern land. A forested continent *has* been found to exist in the southernmost Indian Ocean, at the gateway to Antarctica. It just isn't visible at present.

Approaching Kerguelen Island from the north, the *Erebus* was twice blown offshore in a gale. A razor-like wind cut at the sailors' faces. At last, on May 12, 1840, they skirted a rocky extrusion Ross named "Bligh's Cap," before coming into view of the island itself. Christmas Harbor, on its north coast, had been

Fig. 2.2. Christmas Harbor at Kerguelen Island, which James Cook surveyed in December 1776, on his third voyage around the world.— Engraving after J. Webber (ca. 1785). Wellcome Collection.

made famous by illustrations in Cook's published *Voyages*, and Hooker contemplated its signature archway entrance—a volcanic marvel—with the satisfied wonder of having arrived at a place so often visited in the imagination. He had first thumbed the pages of Cook literally at his father's knee, when the image of the *Resolution*'s men hunting penguins with clubs had inspired boyish nightmares. Now, memory of the geological precision of the drawings returned to him, as the real-world cliffs of Kerguelen loomed through the gray lilt of fog. The entire coast seemed bound by steep precipices, with alternating longitudinal stripes of bright snow and dull rock terminating in a green layer of vegetation sloping into the sea.

After so many weeks being blasted on the open ocean, the *Erebus* and *Terror* looked forward to the sanctuary of the harbor. But they were disappointed. Cook had visited Kerguelen in the relative calm of summer, while the *Erebus* and *Terror* faced pure midwinter fury. The circuitous mountains of the harbor acted like a funnel for the gale-force winds out of the northwest, redoubling their violence. During their two-month stay, a near hurricane blew two days out of three. Despite deployment of every last anchor and cable, the ships were continually laid on their beam ends. Often even the short passage between the ships and the beach could not be attempted by the boats. When they tried to go ashore, they routinely capsized and were dunked in the freezing water. On the beach at Christmas Harbor, the officers conducting magnetic observations at the portable observatory were forced to lie flat on the sand to avoid being blown away. At night, the petrels swarming the rocks moaned in a melancholy chorus with the wind. It was the wildest place Joseph Hooker had ever seen.

His excitement was the greater because he saw in the bright green tufts and bands of brown on the shore the promise of unidentified mosses in abundance. Where the ledges were flat, they were covered in grass. Cook's naturalist, a man named Anderson, had come away in 1775 with only eighteen specimens of plants. Sailing into Christmas Harbor, Hooker thought he could see twice as many from the deck of the *Erebus*. The competitive professional fire—never completely dormant in Hooker—flared up instantly. He would dispose of Mr. Anderson—whom Cook had called "ingenious"—here at Kerguelen Island, just as he would triumph over the French and Americans once Captain Ross had brought them to the pole.

At the south end of the harbor, behind the beach, stood a giant basalt rock that caught the attention of both Hooker and the senior naturalist aboard the *Erebus*, Robert McCormick, as the same rock pictured in Cook's *Voyages*. At the outset of the expedition, Hooker had been wary of the older McCormick, who outranked him and was notorious for his resentful behavior toward Charles Darwin on the recent *Beagle* voyage.

Hooker idolized Darwin (he would later become Darwin's energetic lieutenant in the battle over evolution) and expected similar battle lines to be drawn on the *Erebus*. But McCormick had surprised him with his affability. He was perfectly willing to leave all branches of scientific inquiry to his young colleague except those involving the shooting and stuffing of birds, which he pursued with a vengeance. So when, the day after their arrival in Christmas Harbor, McCormick suggested they investigate the geology of Cook's notable rock, Hooker did not hesitate to join him.

Pulling across the harbor, their boat was slowed by giant kelp, a submarine forest of ghostly fronds grasping at the oars. The kelp leaves lay on the surface, attached by a stem to a pear-shaped bladder on the harbor floor. In the deeper water, two whales—a routine sight in Christmas Harbor—saluted them with flukes raised, and disappeared. The naturalists made their landing on a rock by a beach of black sand, where they were met by a party of penguins. Disgusted by the intruders, the creatures plunged into the surf, emitting a drawn-out honk of protest.

Waterfalls, blown into curling eddies by the wind, surrounded the two men. When the sun appeared, the spray exploded in a sparkling white cloud against the black cliff, showering gently to earth. Soaked by this constant mist, the ground yielded easily beneath their feet, until they were knee deep in gorse-like vegetation. Gazing upward from the valley floor, Hooker was struck by the gorgeousness of the colors created by the moss and lichen adhering to rocks that would otherwise represent only a dull, repulsive mass. Even the Scottish Highlands' rich palette could not compare to the strange beauty of these few species of plant life clinging to existence at this extreme latitude. It was as if the effort for survival itself made them the more beautiful.

When he wasn't killing birds, McCormick's interests lay with geology, while Hooker had emerged as the expedition's dedicated botanist (despite his official rank of assistant surgeon). Ascending Cook's Rock, however, the pair found themselves in a place where rocks and plants merged before their eyes. In the presence of an extruded monolith far younger than the rocky tundra on which it lay, the deep history of Kerguelen Island yawned wide to their astonished contemplation. The great black rock, on closer inspection, consisted of a volcanic greenstone with pebbles embedded, as hard as granite. It was as if it had emerged in a boiling, semifluid state, just yesterday, from the rock beneath. Veins of more recent lava flow ran vertically, like columns. At the base of the rock, loosely embedded, they found something very strange.

McCormick saw them first—bits of charcoal the size of a pipe. He called Hooker over and, though he later swore he had been joking, suggested these were the remains of a fire built by Cook's men or some whalers. They looked around them. Just above their heads, a large fossilized tree, many feet in girth, was stuck in the rock. It required a return visit, and the aid of two strong men from the *Erebus*, to disinter the ancient tree from the rock, and bring it aboard the ship. Before a curious audience in the captain's cabin, they packed the stone tree, with infinite care, into a box the ship's carpenter built specially for it, to survive the years-long transport home.

At the time of the British Antarctic Expedition's visit to Kerguelen Island, its principal claim to fame lay with a unique cabbage. This vegetable sported curvaceous leaves sometimes more than a foot in length, enclosing a white heart—the edible portion—that tasted vaguely of horse radish. Captain Ross ordered the cabbage served with soup, to prevent scurvy. Some in the Admiralty held out wild hopes for the cabbage.

If this nourishing plant could be grown all across the queen's dominions—and quantities of it packaged in tin cans for long voyages—the killer scurvy could be banished from the British fleet forever. But the cabbage proved fickle. Even in the expert hands of William Hooker in the hothouses at Kew—to whom Joseph dutifully sent samples—the Kerguelen cabbage could not be made to grow outside its desolate home soil.

Joseph Hooker detested the antiscorbutic soup, which tasted to him like rotten mustard. For him, the wonder of Kerguelen was not the ubiquitous cabbage but the dozens of other wild plants, mosses, and algae that had somehow escaped notice during Cook's visit in 1776. While the rest of the ship's company was made miserable by the constant gales and rolling of the ship, Hooker beguiled his time dissecting his specimens, examining them under the microscope, then drawing them as accurately as a near-constant ten-foot swell would allow, his legs braced between the wall and the little desk Captain Ross had reserved for him in his cabin.

On those precious days fine enough for shore excursions, not even tramping about in wet clothes in near constant snow could deter him. He found pretty tufted grasses he had never seen or read of. Botanizing alone, he walked along the valley south of the harbor along mossy streams. At intervals, by the little lakes, he found new silk-soft mosses and lichens so plentiful they looked like colorful miniature forests on the rock terraces. Hooker's confidence rose with every fresh discovery added to his bulging note folder, though the cold had adhered the mosses to the rocks so completely that it was difficult to pry loose samples for the Ward's case. How the plants drew nourishment from mere rocks in such a climate defied his ingenuity to explain.

From the moment of the extraordinary find at Cook's Rock, the world of Kerguelen opened before him like a book. Christmas Harbor was a giant crater, only lately immersed by the sea. The island's soaring cliffs—hundreds of layers of igneous rock terraced one upon the other—he now saw were built upon a vast bed of coal, once a dense forest. Even in the silicified lava streams themselves, fossilized trees lay prostrate and exposed to the air, indicating that the Kerguelen forest had regenerated itself, time and again, following successive eruptions from the mountains around him and *beneath* him.

Fossilized algae on the rocks told him that Kerguelen Island—across a span of time he could not begin to calculate—had been alternately submerged and elevated; that forests had grown, thrived, been incinerated by lava, then sprouted again; that the island itself was the stony relic of a far larger body of land, even a continent, on which conifer forests had once stretched beyond the horizon, connected perhaps to the Crozet archipelago to the north, and to the Great Southern Land, if it existed, at the pole.

When he was not botanizing, Joseph Hooker passed his time on Kerguelen Island trekking alone to the summits of the wind-beaten cliffs. He never tired of watching the great surf pound against the precipices of the coast. It was like gazing into time's abyss, to when the clifftop he now sat on was a sandy beach with vanished mountains above, or a forest playground home to a thousand buzzing species now unknown. Volcanoes had done much of the work, it was clear. But one question plagued him. Why the current "desolation" of the island? The now-fossilized Kerguelen forests had risen up under sunny skies, in a mild, inviting climate. And here he was, in AD 1840, forced to sit on the meager floral specimens he had gathered that morning just to thaw them out. As the bitter wind numbed his face, Hooker wondered where the ancient greenhouse nursery of Kerguelen had gone, and all the heat with it.

During one long day's ramble, he descended into a narrow inlet, entirely shut in by towering cliffs and shielded from the constant roar of the wind and surf. The mountain peaks above him were covered in snow, and the waterfalls frozen in mid-descent—as if the perpetual winter of Kerguelen had stopped time itself. The motionless waters of his private bay reflected the black cliffs with mirror-perfect truth. It was eerie to play the role of first explorer in such a place.

Here he came upon the crown jewel of his discoveries: the *limosella aquatica*. This little plant bloomed to full flower, with young fruit, under water sealed by two thick inches of ice: a veritable flower in winter. The ice had been this beauty's greenhouse. Hooker trod warily over the surface of the frozen pond, bent down, and gazed at his prize, displayed through its frozen glass. He broke open the ice and reached down into the freezing murk, where a new wonder unfolded. The close-drawn petals of the *limosella* contained a single bubble of air, generated by the flower itself for its survival. Desolation defied. Hooker dried himself and reached for his notebook. In the remote sub-Antarctic, a tender plant of unimaginable toughness bloomed year to year under the ice without communication with the atmosphere, and without a single human admirer until himself. In this winter waterflower's very existence, the natural order of the world seemed reversed.

In the years following his visit to Kerguelen Island, Hooker's topsy-turvy vision of flowers and ice propelled his botanical reasoning along novel pathways. From those reasonings would emerge the modern science of biogeography. Charles Darwin was the young Hooker's idol, and Captain Ross would later obligingly lead the expedition along the *Beagle's* path to Tierra del Fuego. At that point in the expedition's journey, Hooker had catalogued all flora available to him across the entire southern high latitudes, from South America to New Zealand. From that survey, one extraordinary anomaly stood out. The flora of Kerguelen Island bore a closer relationship to the ecozone of Tierra del Fuego than to Australia and New Zealand, landmasses nearer to it by thirteen hundred miles. In fact, the botany of southernmost Latin America, Antarctica, and the sub-Antarctic islands of the Indian and Southern Oceans appeared to form *a single flora*, despite the near-infinite stretches of ocean between them.

Hooker's only solution to this weird phenomenon was to disagree with Darwin, who was leery of all speculation regarding Earth's ancient continental disorganization. For Darwin, instead of rearranging landmasses on the planet's surface to explain the migration of flora and fauna across vast distances, why not extend the same miraculous powers to the plants and animals themselves, and their power to disperse themselves by wind or sea?

But Hooker had physically traversed those distances across the high southern latitudes for thousands of miles of uninterrupted ocean. For all his respect for Darwin, this experience with the Ross expedition taught him, with embodied conviction, what no theorizing could: that dispersal was impossible, and that only the subaerial existence of "some far more extended body of land" in the ancient past could explain the botany of Kerguelen Island.

Darwin's insight into natural selection, drawn from research in the tropics, launched the intellectual revolution of the Victorian Age, but the great naturalist had a blind spot regarding ice. With his very first scientific paper he had argued, vigorously and wrongly, against ice age theory. And he never adopted the biogeographic theories his friend Joseph Hooker brought back with him from his Antarctic voyage. But our current age of deep ocean drilling has proven Hooker spectacularly right about the deep history of the Southern Ocean—its climate, its botany, and its changing landmasses.

Ancient glaciers carved valleys into the basaltic rock where young Joseph Hooker roamed in the southern winter of 1840. For millions of years, Kerguelen's volcanoes burned and buried its conifer forests at regular intervals. The volcanic deposits, in turn, furnished rich soils for new forest growth out of the ashes. After the forest fires came the ice. Driven by the increased cold of an enlarged Southern Ocean, glaciers plowed the forests, effecting the permanent desolation of Kerguelen Island. When the ice retreated at the outset of the current interglacial period eleven thousand years ago, it exposed the stratigraphic layers of lava streams and petrified forest—half a mile thick—for Hooker's fascinated viewing.

On windswept Cook's Rock, Hooker turned over shards of the ancient forest in his frozen fingers. The inside of the wood was black, shiny, and very hard, with leaf-like striations. Some contained cinnamon-colored crystals; others combined a translucent light-gray pattern with a darker grain. Where a more recent lava stream passed nearby, the ancient wood had turned to charcoal. Everywhere across the site, entire tree trunk sections, some three feet thick, lay compact in their bed of rock.

Once burned, wood becomes resilient to the microbial armies of decay. A carbonized tree will continue to tell the story of its fiery demise for millions of years. Hooker could see the bark and count the rings as if the stone tree had only just been felled by an unknown hand. The irony was not lost on him. On Kerguelen Island, heat and cold were only superficial opposites. The charcoal fragments in his hand belonged to a continuum of fire and ice, forged over time. The fossilized forest represented not only the frozen remnant of a vast ghost continent, but a miraculous foliage dating from the twilight of hothouse Antarctica, before its sudden, definitive leap into the cold.

On the day of the *Erebus*'s departure from Christmas Harbor, Hooker carefully stowed each seed-bearing plant from Kerguelen Island into his Ward's case—a botanical treasure trove he intended would launch his career. The dampness below posed a definite threat, however. After a long discussion with Captain Ross, they agreed to set the collection on deck to escape the humidity; in the event of bad weather, they would rush the case below. A few days out of Christmas Harbor, a violent gale rose up without warning. The hatches had been battened before Hooker even reached the stairs. After an agonizing wait, he rushed on deck. Amid the general mess, he found the waves had broken open the Ward's case and soaked its contents beyond salvage.

The same day, the *Erebus* lost her boatswain overboard. Captain Ross—the speedy French and Americans on his mind—was keen to make up time in their long haul to Hobart Town. Under a full press of sail in buffeting winds, there was never a hope. The man was last seen striking out at the high waves, harassed by a pair of skua gulls with their powerful beaks.

Interlude: The Ghost Continent

In the summer of 1988, the ship *Resolution* of the international Ocean Drilling Program (ODP) set its course for Kerguelen Island for much the same reasons as James Ross did in March 1840: to plumb Earth's secrets in the world's remotest waters. Buried beneath the Kerguelen seafloor lay sedimentary clues to the Hothouse-Icehouse transition and to the Antarctic continent's leading role in that great climate-change drama thirty-four million years ago. Where Cook, Ross, and Hooker found a desolate island outcrop, their explorer successors mapped an entire submarine continent—a sunken remnant of Hothouse Earth—a "France Australe" twice the size of actual France.

The converted oil exploration vessel chartered by the Ocean Drilling Program was known officially as the JOIDES *Resolution*. JOIDES was its bureacratic moniker (tribute to the participating countries in the scientific program), while "Resolution"—named in honor of Cook's vessel—captured the more romantic, naval spirit of modern deep-sea exploration. For the scientists of Leg 120 of the ODP, however, the monthlong passage of the *Resolution* out of Fremantle smacked little of romance. The Roaring Forties greeted them with howling winds and vertigo-inducing seas. Jim Zachos, a graduate student from the University of Rhode Island, was continually sick, barely slept, and couldn't recall when he last saw the sun. Only one heavenly glimpse of the southern lights—the Aurora Australis—relieved the general misery. Meanwhile, deep in the bowels of the *Resolution*, Lamar Hayes, the veteran superintendent (and erstwhile Texas oilman), brainstormed solutions for drilling in fifty-knot winds with a sixteen-foot swell tossing the ship like a bath toy.

Site 748 of the Kerguelen Plateau voyage had been chosen for drilling because its top layer of "recent" sediment—corresponding to the Neogene period that commenced twenty-three million years ago—was very thin, allowing access to more ancient strata and deeper secrets. Zachos, his PhD nearly finished, was scouting a new research project and fresh data. He hoped that a core drawn from the Kerguelen Plateau—one of the largest oceanographic features on Earth—would offer new information about changes in Earth's climate during the forty-million-year stretch between the Neogene and the dinosaurs' demise signaling the end of the Cretaceous period sixty-six million years ago. One "brief" period fascinated him in particular. With a clean core from the Kerguelen plateau, he might date more precisely the boundary dividing the Eocene and Oligocene epochs, the time of drastic climate deterioration that witnessed the transition from Hothouse to Icehouse Earth and helped give rise to our proto-modern oceans, climate, and fauna.

A data-rich core taken from the Antarctic plate region might also solve the mystery of Antarctica's first ice. A feature of the Eocene-Oligocene Transition (EOT) was the steepening of Earth's temperature gradient between the equator and the poles. Hothouse Earth enjoyed uncomplicated, balmy conditions at all latitudes, while our modern Glacial Earth is a far moodier place, with distinct seasons in mid-latitudes and radical differences in climate from the tropics to the poles. But the date of Antarctica's first glaciation—and the mechanism for it—remained elusive. An uneasy consensus gravitated to the creation of a continental ice sheet in the Antarctic fifteen million years ago, with no correspondence to the EOT. But there was no direct physical evidence for that scenario, or any other.

Any prospect that Site 748 could provide a solution to this major scientific question seemed bleak. A day's sail south from Site 747, a distance of 230 nautical miles, aided by GPS, brought the

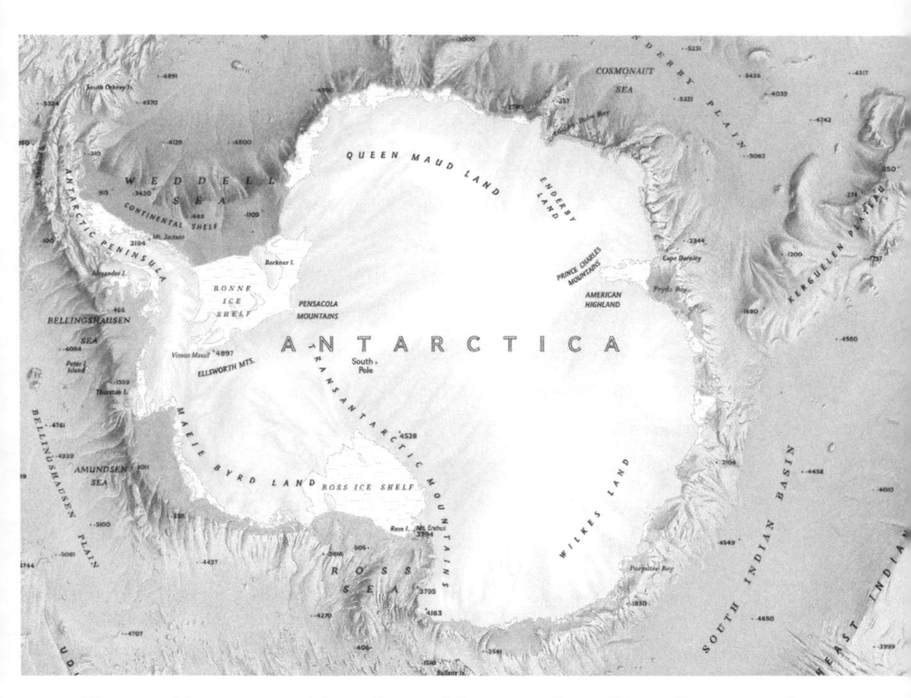

Fig. 2.3. Antarctica, with geology of the ocean floor. Since the 1970s, the international Ocean Drilling Program has drilled cores in the seabed off the coast, seeking to unlock the secrets of Antarctica's climate evolution and its massive ice cap.—National Geographic Image Collection.

Resolution to the designated spot on the evening of March 13, 1988. For two sleepless days, the extreme weather and poor undersea visibility thwarted all attempts to install a full-sized cone at the coring site. On the third day, Lamar Hayes improvised a miniature device to maintain the hole, but at 172 meters beneath the seafloor, the retrieved core collapsed. At 435 meters, Hayes and his team were forced to postpone coring for two hours in order to retrieve a tool that had disappeared down the hole. At 531 meters, Hayes halted drilling again. While the storm raged on, they spent five and a half hours deploying a free-fall funnel to enable a deeper drill.

The second drill pipe reached 388 meters beneath the seafloor before getting stuck at the approximate end of the Cretaceous period. It took an hour to work the pipe free, and another hour and a half to clean it. They had just recommenced drilling when wiring blew on the main cable—another day lost. Then their luck turned. They reached 900 meters, with core recovery better than 80 percent, before the hole began to collapse once more. A float valve failed, flooding the bottom hole assembly. Lamar Hayes had just contrived to decontaminate the pipe when a hose burst, ending operations for good.

They had spent ten days at Site 748—days of continual technical setbacks, little sleep, and a heaving ship. Three days later, the week of his sixtieth birthday, Lamar Hayes suffered a heart attack in the drill room of the *Resolution*. The ship could not have been more remote from assistance. They placed the oilman's body in the refrigerated room for storing core samples and began the long return voyage to Fremantle to take on board a new drill superintendent and complete the mission. Thus the name of Lamar Hayes joined the honor roll of explorers to die in the cause of science at the hostile edge of the world—a list dating back to the first discovery voyages of 1838–42.

Mineralized fragments of the lost Kerguelen forest were scattered through the sedimentary core now lying in state in the laboratory aboard the *Resolution*. Thanks to Joseph Hooker's "stone tree" excavated one hundred and fifty years earlier, their presence came as no surprise to Jim Zachos. Nor were the layers of micro-organic ooze, volcanic basalt, and marine-life remains of tiny teeth, scales, and bones in any way unexpected.

Immediately postretrieval, the Site 748 core had been placed on a long horizontal rack above the drill deck. In that thick, confused muck, seventy-five million years of Earth's history stretched before Zachos and his colleagues. One-half of its length, about four hundred meters, bore witness to the Cretaceous period, the time of the dinosaurs. The rest displayed the sixty million or so years since. The scientists carefully divided the core, labeled each section, and transported them to the lab. With a diamond saw, they bisected each section to create one working sample—for immediate analysis on board ship—and an archival half to be sent to collaborating scientists in laboratories across the world. Finally, they placed both halves of each section of the core into plastic tubes, labeled these, and transferred them to the cold-storage deck, now also a morgue.

The death of their drilling superintendent had demoralized all on board the *Resolution*, and Jim Zachos felt a special responsibility to labor diligently on the last core drilled by Lamar Hayes. Running an eye along the cylinder of preserved muck, his attention was drawn to one fraction of a single section, less than forty centimeters in length. Amid the dark nannofossil ooze—signifying open ocean deposits—he detected a strange debris of sandy quartz, feldspar, and mica. These were foreign minerals that bore no relationship to the volcanic basalt native to the Kerguelen Plateau. Where on Earth had these come from?

Zachos extracted a sample of the quartz debris—the size of sand grains—and hurried them to the microscope. Magnify quartz grains from your favorite beach and they appear rounded and clean, smoothed by geological generations of ocean currents before being deposited on shore. But these grains, extracted at meter 115 of the Site 748 core, were radically different. Some were a translucent grey, others milky looking. But almost all were distinctly angular, with a suite of unusual surface features—arc shapes, step patterns, and dish-shaped fractures—all suggestive of some crushing, external force. The Kerguelen Plateau had sat adjacent to the Antarctic continent almost since its creation one hundred ten million years ago. So, tectonic shifts could not explain the transfer of these quartz grains from the East Antarctic continent to sub-Antarctic Kerguelen, a thousand kilometers away. So too, the coarseness and heft of the grains ruled out their being transported by wind or the ocean current such a great distance.

Ice. Only drifting icebergs explained the presence of quartz and other foreign minerals deep in the submarine body of the Kerguelen Plateau. These tiny souvenirs of the East Antarctic continent had been ground by an immense ice sheet, off-loaded onto an iceberg and transported, as by a raft, to the shores of what was then the terrestrial mini-continent of Kerguelen. Since Antarctica and Kerguelen had not changed their relative position for a hundred million years, this ice sheet, and its iceberg fragments, must have been truly gigantic to survive a one-thousand-kilometer journey across waters warmer than today's Southern Ocean.

Was this the longed-for sign of *first ice*? If so, the discovery at Site 748 had been momentous enough. But when Jim Zachos contemplated the date, he was momentarily staggered. Meter 115 of the 748 core signified 33.6 million years in the past. Antarctica had completely frozen over almost twenty million years before anyone had thought, soon after the Eocene-Oligocene Transition (marked at 120m on the core). If his first reading of the Site 748 Kerguelen core bore out, the climate history of planet Earth would need to be rewritten.

Before the Leg 120 voyage, no consensus could be reached on a start date for the Antarctic ice sheet, when Glacial Earth originated, for lack of physical evidence. Zachos's discovery in the waters off Kerguelen Island changed that. In an expression of bullish triumph rare in scientific publications, Zachos and his fellow researchers dubbed the Site 748 core the "smoking gun" of Antarctic history. They rushed into print with their findings.

The chief scientist on board, Sherwood "Woody" Wise, and his graduate student James Breza, both from Florida State University, described the discovery of the ice-rafted debris. Zachos, meanwhile, took responsibility for isotopic analysis of the oozy sediment contemporaneous to the ice debris at meter 123 of the

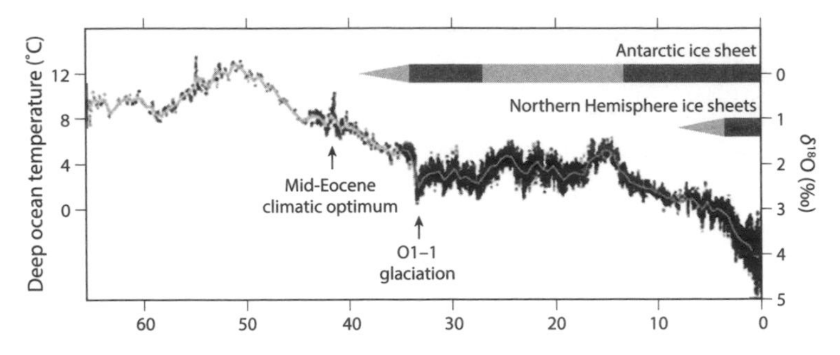

Fig. 2.4. A spike in the heavy oxygen isotope values of deep-sea cores at the Eocene-Oligocene Transition (O1–1) signifies Antarctica's rapid glacial expansion thirty-four million years ago.—Hansen et al., *Philosophical Transactions of the Royal Society* 371 (2013).

core to seek out evidence of colder ocean temperatures coincident with the glacial signal.

A record of ancient sea temperatures is preserved in the isotopic character of fossilized organisms that once lived in that glaciated ocean, secreting calcium carbonate via their shells. The microscopic remains of ocean plankton and foraminifera are a global archive of ocean chemistry and, thus, of climate change. When ocean water evaporates, the lighter oxygen isotope, O-16—being lighter and more volatile—escapes preferentially to the atmosphere, leaving the sea enriched in the heavier O-18. An ice sheet, to form, requires massive precipitation, the so-called snow gun process. An ice sheet thus correlates to an O-18–rich ocean. The O-18 spike, observed globally, had long been interpreted as a signal of rapid Antarctic water surface cooling at the Eocene-Oligocene Transition.

To establish the chronological link between ice-rafted debris on the Kerguelen Plateau and plummeting ocean temperatures at the EOT, Zachos handpicked hundreds of individual fossils from a freeze-dried sample of the 748 core for chemical analysis. Sure enough, the fossils showed an "immediate and profound" increase in the heavier oxygen isotope at the time of the Kerguelen icebergs. At meter 115—33.6 million years ago—the missing O-16 was already locked in the ballooning Antarctic ice sheet.

Major scientific findings—especially those deserving the title "smoking gun"—require multiple, converging lines of evidence. To complete the triumph of the 748 team, Sherwood Wise's colleague at Florida State, Wuchang Wei, analyzed microfossils in the same section of the core and confirmed an abrupt increase in the numbers of tiny cool-water creatures—plankton and single-celled foraminifera—present in the early Oligocene waters of Kerguelen.

In the findings of Zachos and Wei, the confirmation of drastic Southern Ocean cooling at the EOT tallied with well-established data for the rest of the world's oceans. The breakthrough lay in identification of a trigger for the global temperature crash. In the new Antarctic history uncovered at Site 748 off Kerguelen Island, polar glaciation was not a byproduct of global cooling—a gradual process by which a continental ice sheet developed fifteen million years ago. Rather, the much-earlier initiation of an Antarctic ice sheet had itself driven the icehouse cooling of the world at the beginning of the Oligocene—by spinning up the oceans, revolutionizing their chemistry and biota, steepening the temperature gradient from tropic to pole, and fueling winds and storms.

The literally epochal event of Antarctica's freezing over now acquired its own technical name: the Oi-1 Glaciation (the "Oi" stands for the oxygen isotope anomaly that is the proxy signal for glaciation). As Zachos explained in a 1996 paper, "Oi-1 was an extreme but transient climate." It was not "a random event in time," but rather a highly unstable transition "between two quasi-stable climate modes," namely Hothouse Earth and Icehouse Earth. "Transient climates of this magnitude," he concluded, "are rare in Earth history." At the Oi-1 Glacial Maximum, deepwater temperatures had cooled from their tropical Eocene highs to a near present-day chill, while Antarctic ice volume metastasized from a few isolated frozen lakes at high elevation to a glacial sheet a quarter larger than the size of the current ice cap. Sea levels dropped nearly a hundred meters. Global hydrology entered a new age. A template for modern Earth was set.

For geologists, the establishment of the Oi-1 Glaciation required a new narrative of Earth history. The asteroid that struck near Mexico sixty-six million years ago might have cleared the way for mammals like ourselves, but the Oi-1 Glaciation was the most critical Earth system change since. Even the iconic "ice ages," too, appeared differently—less as a planetary thermal reorganization in themselves than as a recent, northern hemispheric iteration of a global polar ice regime initiated thirty-four million years ago in the newly opened Southern Ocean.

One question remained: what had caused Antarctica to ice over in the first place? Ten years after the momentous ODP drilling venture in the southern Indian Ocean, which unearthed proof of the origins of the Antarctic ice sheet and the evolution of Glacial Earth, the JOIDES *Resolution*—Leg 189—returned to the waters off Kerguelen Island. Its mission, this time, was to better understand the volcanic character and origin of the "Large Igneous Province"—the sunken Kerguelen Plateau—of which Kerguelen Island was among the few subaerial remnants. The massive volcanic convulsion that produced the Kerguelen plateau is characteristic of terrestrial formation on the moon, Venus, and Mars, as well as Earth. This 480,000-square-mile plateau—whose existence Joseph Hooker guessed at—is as large as any of its kind on our planet.

The magmatic eruption sites that produce igneous provinces, actually landforms, are called hotspots, a word that does not do justice to this world-shaping phenomenon. Hotspots can be four hundred kilometers in radius and be active for tens of millions of years. The Kerguelen hotspot emerged during the Cretaceous period over one hundred million years ago—a far more volcanically active time in Earth's history than the present—when its fiery 300°C plume concentrated the energy of Earth's mantle on young Indian Ocean crust. The heat budget and material output of the Kerguelen hotspot far surpasses any volcanic event witnessed by humans; it lies outside any ordinary imagining. As first the slow-moving Indian plate, then the Antarctic plate, rotated across the active hotspot plume, vast quantities of magma were emplaced as new land. The Kerguelen hotspot, in its hundred-million-year history, created a continent's worth of mountains, valleys, and plains, stretching from the Indian Ocean east of Sri Lanka to the shores of Antarctica.

The 1998 ODP cores contained fossilized wood fragments, fronds, leaves, seeds and spores from these halcyon days of the Kerguelen landmass, when it hosted a dense tree canopy and giant ferns carpeted the forest floor. The changes in sediment encasing these fossils—from volcanic rocks, to intertidal sands, to the "ooze" of open ocean—told the vivid history of the Kerguelen micro-continent's gradual subsidence. Over tens of millions of years, the volcanic land cooled and slowly sank beneath the waves of the newly formed Southern Ocean, entombing its forests and the bones of its creatures and leaving only desolate Kerguelen Island, and a handful of other pinpricks of land in the sub-Antarctic, as clues to its once grandiose, sunlit existence.

The 1998 mission likewise added a fascinating chapter to the history of our understanding of Kerguelen Island itself, a scientific history that began with the visionary speculations of Joseph Hooker in the southern winter of 1840. While the earliest of the Kerguelen hotspot's output dates back one hundred thirty million years—placing it among the longest-lasting known volcanic sites on Earth—Kerguelen Island emerged relatively late in that hotspot's history, in the late Eocene, about forty million years ago. Where other tracts of the plateau enjoyed perhaps fifty million years as bona fide landmasses of Hothouse Earth, the island forests stretched toward the sky only a few million years at most, before its once-fertile terrain sank beneath the waves, the climate froze over, and Kerguelen's long era of desolation began.

‡ 3 ‡

In the Land of Fire

New Year's morning of 1838, the year prior to Joseph Hooker's productive residence on Kerguelen Island, found the front-running French ships *Astrolabe* and *Zélée* bobbing at anchor five thousand miles to the west, in the Straits of Magellan, the fabled southern concourse between the Atlantic and Pacific Oceans. Here, in Patagonia, below 46° south, glacier tongues descend from the Andean peaks into the labyrinthine fjords of Tierra del Fuego—the largest mass of continental ice in the Southern Hemisphere outside Antarctica.

In a cutting wind on deck, Dumont D'Urville presented each of his officers with a silver medallion. It depicted their commander's Roman profile on one side and an image of the *Astrolabe* and *Zélée* breasting the open sea on the other. An inscription read, "In Honor of the French Expedition to the South Pole, by Royal Decree, etc., etc. January 1st, 1838." Their orders were to sail south from Tierra del Fuego across the roughest seas on Earth, then to keep going until what was there, or the nothing that was there, stopped, engulfed, or destroyed them.

That is, if the Americans did not win the honor first. At Rio, an American merchant captain boasted to D'Urville on good authority that the United States Exploring Expedition, so long in preparation, had sailed last October bound for Antarctica, commanded by a determined man named Catesby Jones.

D'Urville could not know that this information was entirely false. As far as he knew, the Americans had reached Antarctic waters weeks ago. There would be no royal prize, no glory, if the Yankees beat them to the pole.

On their arrival in Patagonia, D'Urville climbed to the lip of the glacier above the beach at Fortescue Bay, bringing his reluctant steward with him. Far below, they could make out the tiny shapes of the *Astrolabe* and *Zélée*, like twin cockleshells on the glimmering water. Originally designed to transport horses, the three-masted corvettes' capacious holds were well suited to a years-long mission of discovery. To battle the polar ice, each had their prows stiffened with bronze, and hulls double-sheathed in copper. Her open gunports, on the other hand, were conspicuously vulnerable to a true Southern Ocean gale. And though he had sailed twice around the world in the *Astrolabe*, D'Urville could not love the ship, whose dingy furnishings would have shamed an American whaler. The French commander was keenly aware of the advantages in funding and equipment enjoyed by his rivals. On the other hand, he knew the South Seas better than any of them.

The glacier's crest looming above their heads was covered in half-melted snow. D'Urville and his shivering steward could hear the glacier's fast-flowing interior currents as it underwent its seasonal melt. When the wind off the ice picked up, D'Urville lost all feeling in his fingers. He had forgotten his heavy coat and a strange weakness overcame his limbs. They hurried back down to the beach. The summer, such as it was, was already in retreat. It was time to take the ships to the ice.

New Year's Day spent in a wild Patagonian forest lifted their spirits. On a sandy bank by a glacial stream they arranged their feast: a goose roasted on a spit, with bottles of wine and champagne to ring in 1838. The officers took the opportunity to sound out the commander's intentions. How far to the south would he take them: 76°? 78°? Would they beat the

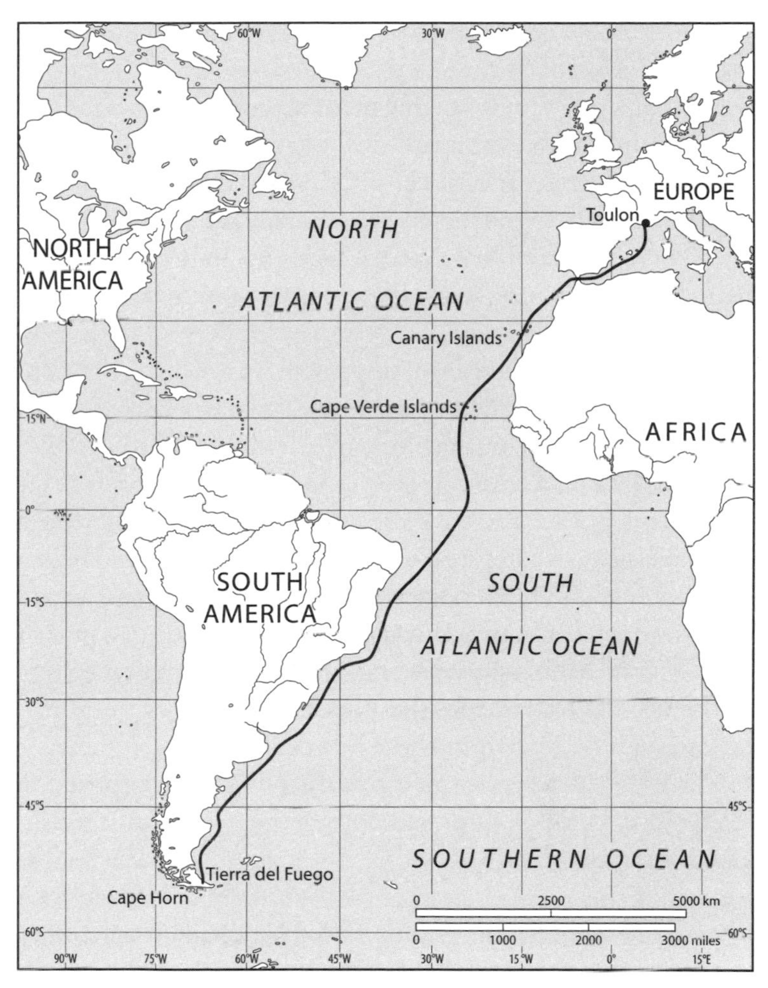

Fig. 3.1. The transatlantic route of the French *Astrolabe* and *Zélée* from Toulon to Tierra del Fuego, September–December 1837.

English whaler Weddell's record southing? "If Weddell's account of the ice is correct," D'Urville told them with a smile, "we will not turn back before 82° south!" D'Urville was a bulky, shabby man, with a grand face as impassive as marble. The officers stared now at their laughing captain, who had not evinced a single emotion across eight thousand miles of Atlantic Ocean.

Back on deck, they found the men rummaging through their new Antarctic stores: enormous sheepskin boots from Newfoundland, greatcoats, thick vests, and flannel underwear. Lambswool gloves completed the costume, so that every man on board appeared to have grown twice his natural size. Then they danced a little on deck, like an ocean-going race of Patagonian giants.

Their final act before departing south was to bring letters to the makeshift box nailed to a solitary tree near Port Famine, so named for the Spanish colony that had starved to death in the Magellan Straits two centuries before. If they were never to return from the ice—as François Arago had sworn would happen—it was a comfort to know their families would have a final expression of love to remember them by. By chance, an American whaler passed through Port Famine only a month later. Via the haphazard yet wonderfully reliable postal system of the sea, the letters from the literate members of the *Astrolabe* and *Zélée*, sent from the wilds of Patagonia in January, reached France in June 1838, where the public was surprised to learn old D'Urville's ships had not yet been crushed to smithereens in the ice.

Their first evening in the Straits of Magellan had brought every last man on deck to witness the sun setting behind the glacial peaks of the Andes. The dark clouds at the summit filtered an orgy of color—flaming purple, orange, and red—in perfect imitation of an erupting volcano. Where the sun disappeared, the lurid sky evolved by degrees to a filmy yellow radiance that brought the southerly mountains of Tierra del Fuego into spectacular relief. Beneath these the islands of the straits formed a jagged chaos, as if only just deposited by some work of planetary violence. None of the gentler spirits of nature prevailed here. But with the sun at its lowest point, they could perceive each mountain, separated by a dense, forested valley, succeeding to the next perfect alpine image, and so to the horizon, which seemed to mark the very edge of the world—as, in a sense, it was.

Just to the west of where they stood, in 1520, the explorer Ferdinand Magellan had faced down a mutiny. His men pointed to the massive, crenellated face of Cape Froward and accused him of leading them through the Gates of Hell. Like Magellan, D'Urville's men were contemplating the southernmost vegetation on Earth. The theatrical scene concluded with gale-force winds from the west howling under the cliffs and a sudden onslaught of snow and hail from a black sky. Renaissance mapmakers imagined Tierra del Fuego as the northern tip of Terra Australis Incognita: a geographic error, but meteorologically accurate. Relentless, frigid westerly winds link Tierra del Fuego to an Antarctic climate regime distinct from latitudes north. This was particularly the case during the last ice age, when the gloomy valleys and channels of the present-day archipelago were chiseled out by glaciers.

Every officer aboard the *Astrolabe* and *Zélée* burned with desire to meet the local Patagonians, who enjoyed mythic status in Europe as giants and man-eaters. The reality proved grimmer and more complicated. On the silver beaches of the Magellan Straits, the French found ceremonial graves dug in the sand filled with the burnt remains of animals, clothes made of hide, and a warrior's weapons. The body itself was seated upright, as if alert to the call to the next life beyond the Andes mountains. When the Patagonian died, his kinsmen killed his dogs and horse, then filled the hole with his meager worldly possessions. During the brief inferno, the smell of burnt flesh and animal hide filled the beach. For a man of true merit, the women of the family wailed inconsolably for three days, then promptly forgot him. The deceased had been duly transported, via smoke, to the transalpine paradise. It would be superfluous to mention him again.

On the wooded northern shores of the Magellan Straits, local family groups set up temporary tents with branches and animal skin. Naked children chewed berries and brawled on the beach, while the women—a distinctive red line painted beneath their eyes—waded chest deep in the frigid water searching out mollusks with their feet. Salted guanaco meat hung in long strips from cords strung between the tents. The Patagonian men, who hunted the llama-like guanaco, rode horses brought by the Spanish, and smeared themselves head to toe in animal fat for insulation against the cold. They showed no signs of being overawed at the French presence; Europeans had been visitors to Patagonia for three centuries. Among those wearing the native mantle of guanaco hide, clasped at the neck, was a Fuegian dressed in full European attire, complete with breeches, vest, and frock coat. Tobacco, alcohol, and sugar had already corrupted the Patagonians' health, and while they looked scornfully on an offer of glass beads, they could be induced to exchange a luxurious guanaco fur for a single biscuit. But steel—knives and swords—was valued most of all, as a priceless import to their stone-age existence.

D'Urville was surprised to find among the cold-climate natives two actual Europeans, castaways of the seal trade. The Patagonians had kindly taken them in, but the extreme cold and raw diet had traumatized the sailors, who wept openly on their rescue by the *Astrolabe*. With their help, D'Urville began work on a French-Patagonian dictionary—with its several dozen words for "cold"—and conducted stilted conversation with a distinguished local he invited to dine. This chief spoke an amalgam of Spanish and French, with which he made sure to compliment his host: "Ingles no good, American no good, French bueno." After eating his fill, he asked to be returned to shore and was given a long knife as a token of goodwill. These Patagonians—the robust hunters of the north shore—impressed D'Urville. But their southern neighbors, from Tierra del Fuego proper, led a more wretched existence. These were fishermen, dependent on their canoes and the marine kelp ecosystem—a smaller race, proportionally less forthcoming, who spoke an entirely different language. How had two apparently separate races, with divergent lifestyles and technologies, come to inhabit the same desolate place?

Dumont D'Urville, over the course of his three voyages, encountered more indigenous peoples of the Southern Hemisphere than any other nineteenth-century European. Despite his Eurocentric prejudices, he was a gifted anthropologist. It was D'Urville who first recorded linguistic differences among the South Sea regions of Polynesia, Melanesia, and Micronesia—categories still in use today. He even brought a phrenologist with him aboard the *Astrolabe*, a man named Pierre Dumoutier, in the hope of collecting skulls and plaster life masks for "scientific" examination. But Dumoutier was disappointed in the Patagonians, who wanted nothing to do with skull collecting or having their faces covered in plaster (he slept with the local women instead).

A question tugged at D'Urville, as it did every European to visit Tierra del Fuego: how did the Patagonians come to inhabit the frozen tip of South America, and how did they survive the extreme conditions there? In the words of Charles Darwin, a recent visitor to the region: "What could have tempted, or what change compelled a tribe of men to leave the fine regions of the north, to travel down the Cordillera or backbone of America, to invent and build canoes, and then to enter one of the most inhospitable countries within the limits of the world?"

Fig. 3.2. The indigenous people of Tierra del Fuego were an object of fascination, and morbid fantasies, for nineteenth-century Europeans. Driven to extinction soon after their encounter with the Victorian polar explorers, they are the only known human community to have lived below 55° south.—Charles Darwin, *Journal of Researches into the Geology and Natural History of the Various Countries Visited by HMS Beagle* (1839). Rare Book and Manuscript Library, University of Illinois at Urbana-Champaign.

At Cape Remarkable, the scientific gentlemen of the *Astrolabe* spread out in search of the fossilized mollusks their legendary compatriot Bougainville had claimed to discover high up the cliffs, alongside giant bones of vanished species. But these only deepened the mystery. The seas had risen. Giant mammals had lived and died, outlived by the Patagonians. But when and in what order, and what had brought that prior world to its end?

Interlude: The American Climate Warriors

Charles Darwin's botanical research in Tierra del Fuego was much on Joseph Hooker's mind as he piloted the *Erebus*'s boat to shore at Kerguelen Island. The fronds and tendrils of a submarine jungle clogged the surface of Christmas Harbor, so that two men working the oars could barely make progress. Hooker recognized one of the seaweeds from European waters: the *Macrocystis pyrifera*, so-called giant kelp. But its companion marine weed—leathery, tubular, and buoyed by hefty bladders—was a stranger to him. He found an impressive example of the new kelp washed up on the beach, sixty feet long, that he could barely cram onto the boat.

Back on board the *Erebus*, Hooker consulted his botanical library, daring to hope the honor of discovery would fall to him. In the just-published *Voyage of the Beagle*, Darwin's description of giant kelp in Patagonian waterways had made a deep impression. For Darwin, only the terrestrial jungles of the tropics could be compared to the "great aquatic forests" of the southernmost inhabited land on Earth. With the aid of its disc-shaped foot, kelp attached itself to every available rock in the Magellan Straits and on the glacial Chilean coast, where it withstood the relentless surf or, if detached, lugged its fraternal boulder with it to shore.

When he shook out the entangled roots of *Macrocystis pyrifera* on the deck of the *Beagle*, Darwin discovered a living zoo: from the microorganic coral incrustations decorating each frond to innumerable polyps, algae, mollusks, and crustacea to cuttlefish, crabs, starfish, sea eggs, crawling nereids, and schools of small fish of stunning variety. This marine banquet in turn fed the ever present birds—the cormorants, albatrosses, and petrels that circled the *Beagle*—as well as otters, seals, and dolphins that crowded the bleak channels and islands of Tierra del Fuego. Contemplating this colorful menagerie, Darwin conjectured that nowhere else in the world did so many species depend for their existence on a single plant. These included the maritime peoples of Tierra del Fuego, who had adapted somehow to the sub-Antarctic cold and sustained themselves on fish and seal flesh garnished by dried tidbits of the seaweed itself. Were the kelp to be destroyed, a mass extinction, including humans, would inevitably follow.

Darwin had not mentioned Hooker's mystery kelp, lying in a confused heap under the stern window of the *Erebus* captain's cabin. Unfortunately, however, Hooker's consultation of the *Dictionnaire Classique de l'Histoire Naturelle* confirmed that a French botanist had preempted both him and Darwin. And no ordinary Frenchman it was. Jules-Sébastien Dumont D'Urville, present commander of the French Antarctic Expedition, had identified the indigenous giant seaweed during the first of his three Southern Ocean voyages on the beaches of the Falkland Islands, east of Tierra del Fuego. Taking his cue from the native Fuegians' diet, D'Urville had turned the seaweed into a tasty soup (you can find the dried kelp at Chilean markets today, tied in bundles like asparagus, and sold as "cochayuyo").

D'Urville had transported his tentacular specimen back to Paris, where the botanist Bory de Saint-Vincent bestowed on it the name *Durvillaea antarctica*, or "Durvillaea utile," because Dumont D'Urville had performed such very useful service to the natural sciences in his South Sea voyages. Perhaps other resemblances between the famous French explorer and the bull kelp of the Antarctic Ocean helped inspire the name. According to its official classification, the character of *Durvillaea antarctica* was "robust, buoyant . . . and capable of rafting vast distances," a persuasive description of its namesake.

When James Ross brought the *Erebus* and *Terror* east from Kerguelen Island to the British colonies of Tasmania and New Zealand in August 1840, his first request was for news of D'Urville's expedition. Meanwhile, Hooker found evidence of their French rival everywhere along the rocky coasts in the tenacious presence of *Durvillaea antarctica.* Even when the British expedition set sail south from Hobart into the polar unknown, the bull kelp named for their nemesis, like fingers stretching from the deep, kept company with them on the open sea. Only when they entered the realm of ice did the circumpolar *Durvillaea antarctica* concede its natural limit. In his notebooks, Hooker described the kelp, admiringly, as the southernmost remnant of vegetable life; it marked the northern border of the ice. Beyond D'Urville's bull kelp lay a different world, with an unfamiliar purity of air, untainted by rot.

D'Urville's kelp contains vital historical climate information, also. Genetic analysis of *Durvillaea antarctica* has revealed a striking homogeneity among species samples at all meridians. This suggests the bull kelp's current range represents colonization since the last ice age and marks the historical reach of that glacial maximum significantly farther north than previously thought.

Later, in 1843—their feats of Antarctic discovery behind them—the *Erebus* and *Terror* dropped anchor in the Falkland Islands where, two decades earlier, Dumont D'Urville had first trod among washed-up remains of his eponymous kelp. Just westward lay the Magellan Straits, from whence the *Astrolabe* and *Zélée* had sailed southward five years before with the south polar prize at their fingertips—so the Frenchmen thought.

While in the Falklands, the *Erebus* received news of D'Urville's shocking death. Ross and Hooker were dismayed. Neither had wished their French counterpart outright victory in the Antarctic race, but D'Urville's was an irreplaceable loss to scientific

Fig. 3.3. *Delesseria hookeri*, a Southern Ocean seaweed discovered by, and named for, Joseph Hooker. The Victorian expeditions made pioneering contributions to Antarctic marine biology.—Joseph Hooker, *The Botany of the Antarctic Voyage* (1847). Rare Book and Manuscript Library, University of Illinois at Urbana-Champaign.

exploration, as well as a reminder of their own tenuous grip on life (and fame), so far from their home ports. The *Durvillaea antarctica*, they knew from their own experience, was a notoriously difficult plant to study and store. When dissected, it turned an unnatural black. Transported across the oceans, it shrank to a mangled caricature of itself. Now that the great man was dead, D'Urville's enemies would no doubt do all in their power to script a similar fate for his reputation.

The seaweed *Durvillaea antarctica*—and, by extension, D'Urville himself—played a conspicuous cameo role in the solution to the mystery of the Patagonians' origins, which had so occupied the French explorers of 1838.

In the summer of 1977, on a field trip in northern Patagonia, the American archaeologist Tom Dillehay made a stunning discovery. Digging by a creek in a nondescript scrubland called Monte Verde, he came upon the remains of an ancient camp. A full excavation uncovered the trace wooden foundations of no fewer than twelve huts, plus one larger structure designed for tool manufacture and perhaps as an infirmary. In the large hut, Dillehay found gnawed bones, spear points, grinding tools and, hauntingly, a human footprint in the sand. The ancestral Patagonians had erected their domestic quarters using branches from the beech trees of a long-gone temperate forest, then covered them with the hides of vanished ice age species, including mastodons, saber-toothed cats, and giant sloths.

Fire pits indicated where the Monte Verdeans cooked their food. Grinding stones helped fashion their spear points for hunting, while, scattered on the excavated floor of the large hut, Dillehay uncovered the fossilized remains of more than twenty medicinal plants, including *Durvillaea antarctica*. Because seaweed is short-lived, the *Durvillaea* fossil provided the most precise available date for human occupation of the site. Radiocarbon analysis of charcoal, worked wooden artifacts, and the leftover bones of a mastodon meal confirmed the presence of ice age hunter-gatherers at Monte Verde fourteen and a half thousand years ago, at least a thousand years prior to any existing archaeological evidence for human colonization of the Americas, North or South.

When modern humans migrated out of Africa around sixty thousand years ago, they dispersed rapidly through Asia, and thence to Europe. Seafaring peoples hopscotched across the Pacific and reached Australia via a land bridge from Indonesia. But for tens of thousands of years, America remained terra incognita, barricaded by ice. In the 1930s, a site uncovered in New Mexico, called Clovis, suggested that early human hunters had penetrated south along an ice-free corridor no earlier than when the northern glaciers commenced their retreat eleven thousand years ago. Once in the Great Plains, these first explorers discovered a rich array of megafauna—mammoths and mastodons—entirely unprepared for human aggression, and gorged themselves accordingly. It made for an arresting narrative: the First Americans were marauding big-game hunters who braved the ice and quickly established themselves as all-conquering apex predators of the New World.

Dillehay's discoveries at Monte Verde in Chile shattered this theory and threw American archaeology into a period of bitter turmoil. With the emergence of the Monte Verdean foragers, an entire new history of human colonization of the Americas was required. Allowing some thousands of years for these pioneers to make their way from Siberia across the Beringia land bridge to the tip of South America—adjacent to the opposite pole from where they began—necessitated a first entry date of approximately fifteen thousand years before present. At this time, the glaciers of the last ice age were still near their maximum, and no ice-free corridor existed through the western plains of modern Canada. That left the unglaciated Pacific coast as the only possible migratory route. Instead of butchering their way into the heartland as big-game hunters, the First Americans were beachcombers—maritime opportunists who rafted the innumerable estuarine waterways of ancient California, living off the smorgasbord of sea creatures inhabiting the so-called kelp highway of the west coast.

Unlike the inland corridor, this marine route for the First Americans was linear, unobstructed, and entirely at sea level. Giant seaweed forests—including *Durvillaea antarctica* and *Macrocystis pyrifera*—hosted protein-rich populations of giant sea bass, cod, rockfish, sea urchins, abalones, and mussels. A now-extinct species of sea otter offered variety to the First Americans' palate. The seaweed fragment of *Durvillaea antarctica* that Dillehay had found at Monte Verde signified more than an iodine-rich medicinal supplement for ancient hunter-gatherers. It was a clue to the entire history of American colonization, whereby the first humans had followed the rich food source of the kelp highway along the Pacific coast from Alaska to Chile. A rare clue, necessarily. The vast majority of the American kelp highway and its way stations—traveled when sea levels were up to a hundred meters lower than today—is now submerged beneath the Pacific Ocean.

Climate change had opened the highway south, but weather did not always cooperate with the First Americans. In fact, wild climate fluctuations of the Late Pleistocene epoch—some slingshot glaciations unfolding in the course of a human lifetime—made for a stressed, low-density community of humans who could boast of little more than heroic extreme-weather resilience. Thousands of years before Dumont D'Urville navigated the chill waters of the Straits of Magellan, the ancestral Fuegians had ventured into deep South American terrain released by rising temperatures and retreating glaciers. Monte Verde was one such frontier domain. But then, about fourteen thousand five hundred years before present, the climate pendulum swung back, catching the pioneers by surprise. The so-called Antarctic Cold Reversal—when average temperatures fell up to 6°C below present day—lasted two millennia, during which the ancient Americans huddled for survival in caves. When temperatures warmed again, the largest southern glacier outside Antarctica melted, flooding the Straits of Magellan, and cutting off the southernmost adventurers from the mainland. The cultural division of Patagonian peoples into mainland hunters and coastal canoeists—a mystery to the Victorians Darwin and D'Urville—originated with this early Holocene warming eight thousand years ago.

As controversy over the new coastal migration theory raged through the 1990s, another discovery—at the polar opposite end of Earth—added to the glacial mystique of the First Americans. Near the River Yana, at 71° north, above the Arctic Circle in Siberia, Russian scientists uncovered a hunting camp some forty thousand years old. The decorated horn of an extinct rhinoceros and detachable spear shafts anticipated the Clovis peoples of North America many millennia in the future. Meanwhile, genetic analysis concluded that all modern indigenous Americans can trace their ancestry to a single community in this very region of Siberia, perhaps a mere thousand strong.

Between Yana River and Monte Verde, transcontinental traces had now emerged of what was surely the Greatest Journey of All Time. Over the course of twenty thousand years, a small, intrepid human group somehow adapted to Arctic tundra conditions in the midst of an ice age, surviving −40°C temperatures *without firewood* in animal skin tents. No modern-day analogy exists for these climate warriors, which means we can't honestly guess how they coped. We then lose track of the Yana people for several hundred generations before their putative descendants' reemergence as fishermen-navigators, pushing southward along the Pacific Coast into warmer climes, all the way to modern Chile. Then the glaciers rebounded, and a small frontier group was trapped south of latitude 55° and had to adapt all over again to polar extremes.

Darwin's question—how had the Patagonians come?—has been answered by the kelp highway theory. The marine cornucopia hosted by *Durvillaea antarctica* and *Macrocystis pyrifera*—seaweeds that Dumont D'Urville and Joseph Hooker recorded all across the sub-Antarctic realm—had lured the prehistoric Americans ever farther southward. Since then, Patagonian lifeways remained remarkably static: no leap to agriculture or cattle-rearing was possible in the sub-Antarctic landscape. But how the marooned Fuegians managed to survive the polar cold at the world's end for thousands of years—as their Yana forebears had done north of the Arctic Circle—continues to baffle and amaze.

Patagonian skulls provide an incomplete answer. D'Urville's phrenologist, Pierre Dumoutier, failed to gather the native craniums he craved. But latter-day skull scientists—cranial morphologists—have identified cold-adaptive traits common to Fuegians and the Inuit of the Canadian Arctic. Freezing air is deadly to human lungs; hence the nostrils plays a vital role in warming the breath before it bathes the tender membranes of the chest. Adapted to polar temperatures, the nasal cavities of the Fuegians and Inuit are the most capacious of all modern humans, allowing for maximum air turbulence and residence time for each inhalation. Fuegian skeletal remains nevertheless show marked evidence of temperature stress: these include chronic inflammation of bone tissue, arthritis of the back, and overworked masticating jaws, as well as osteomyelitis, in which bones become infected and "die"—all indicators of cold trauma. The Fuegians survived, but they suffered. Little wonder Patagonian oral history revolves around myths of glaciation and angry deities of freezing rain and storms.

For all their epic climate resilience, however, there was no surviving European invasion. Five years after D'Urville's visit to the Straits of Magellan, the Chilean government sponsored its first settlements in Patagonia, setting the indigenous communities on a fast track to extinction. No biological remains exist to determine whether any other micro-evolutionary adaptations had enabled the Patagonians to endure for thousands of years south of latitude 55°. A fat-intense diet—dependent on the abundant seal population of the straits—must have been indispensable to their survival. Thus it's possible that the emaciated condition of the maritime Fuegians, as D'Urville encountered them, was due to the already significant inroads of the European fishing industry on Southern Ocean seal populations. One species of fur seal—the *Arctocephalus gazella*—had already been wiped out on the neighboring South Shetland and South Georgia islands. Their Fuegian descendants who were not killed by measles, alcohol, or guns froze to death from lack of seal meat insulation.

European awe of the Patagonians resulted from a double sense of difference. They wondered at the Patagonian adaptation to the cold, while being humbled by their own intolerance for sub-Antarctic extremes. Hundreds of Spanish colonists had died of exposure and starvation in the Magellan Straits in 1583. They ate seaweed and mussels in imitation of the locals, but not enough guanaco or seal to keep out the cold. A decade later, Englishman Thomas Cavendish took the *Roebuck* into the same waters, where his men expired from "the extremity of frost and snow" at the rate of eight or nine a day. Cavendish deposited his hypothermic invalids on the beaches to die, while the remnant crew threatened mutiny from "their ardent desire of being out of the cold."

This physical intolerance for sub-Antarctic latitudes translated into great mental stress for all Europeans who ventured there, beginning with Magellan's mutineers, whom the explorer felt compelled to hang, draw, and quarter on deck as an example to the rest of his men to bear the cold without complaint. Francis Drake dealt similarly with a rebel officer on his own voyage around Cape Horn. Given the option to be marooned, face trial in England, or be beheaded on the spot, the shivering gentleman chose the axe. More recently, during the *Beagle*'s first voyage to Patagonia in

1827–28, her captain, Pringle Stokes, became so despondent at the "dreariness and utter desolation" of the Straits of Magellan that he shot himself. This was a place, he wrote in his diary of Patagonian despair, where "the soul of man dies in him."

A decade later, in January 1838, Dumont D'Urville and the men of the *Astrolabe* and *Zélée* were no better prepared for the Antarctic climate. A midsummer mild spell had spared them while in the Straits of Magellan, but now they were to turn their ships southward once again, this time across the fierce Drake Passage and into iceberg-filled seas. From the land of fire to the land of ice. To D'Urville, who knew all the Patagonian horror stories by heart, it must have seemed something like a suicide mission.

PART TWO

⚓

Trials

‡ 4 ‡

D'Urville Battles the Pack

Two weeks after they set sail from Tierra del Fuego, the *Astrolabe*'s lookout spotted breakers on the horizon: a telltale sign of approaching ice. They soon found themselves drifting among blocks the size of rowboats. Meanwhile, the captain's giant kelp, *Durvillaea antarctica*, had disappeared from the surface waters. They were entering an odorless, unvegetated, alien world.

As the last of the ocean warmth ebbed away, the mist parted, and a scene of melancholy beauty emerged: immense blocks of ice with sheer walls, here an arch hollowed out by the waves, there a perfectly rounded scoop. The men crowded the side to watch the floes pass. Bubbles rose to the sea's surface like frozen rivers melting at the end of winter. Stinging spray broke from each crest of the swell. When the wind rose, the sea pounded the ice. The waves unfurled across the ice rafts with a violent flop and a crash, like over a coral reef.

The *Astrolabes*'s attention was soon drawn away from this petty flotsam to a giant ice prism off the bow, a hundred times the size, surrounded by fog. Gusts of wind shifted about its Olympian peak, breaking the clouds. Its flanks glinted in the sunlight. Dumoulin, the *Astrolabe*'s navigator, estimated its length at two hundred feet, though the men swore among themselves it was twice as big. It towered above the ship's mainmast. How many winters, snow on snow, had passed to create this monster? The ice mountain withdrew into the fog, casting

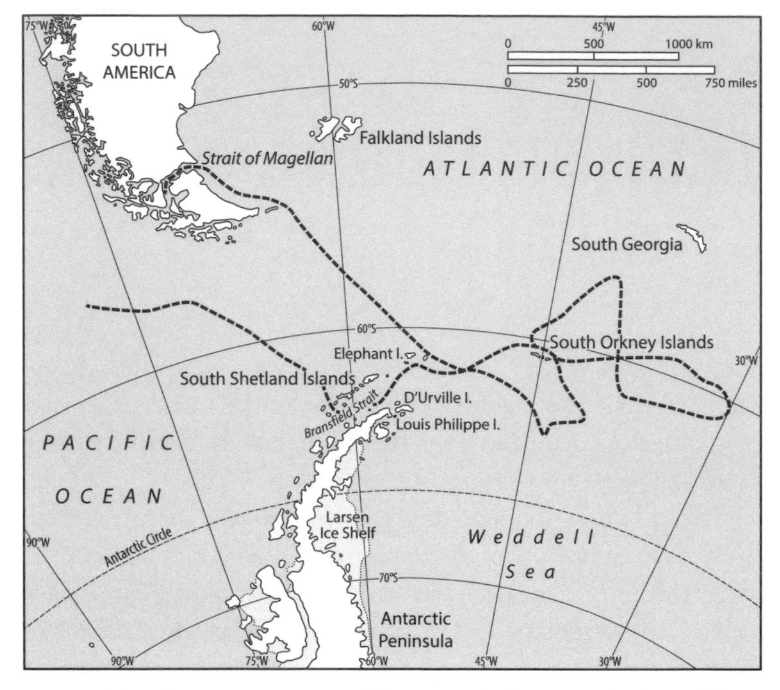

Fig. 4.1. The dangerous vagaries of the ice pack determined D'Urville's haphazard course in the Weddell Sea in the southern summer of 1838.

a fugitive shadow on the water, while squalls of melting snow blinded them. D'Urville ordered double rations for the men, to mark their first sighting of Antarctic ice. He also quietly doubled the watch to ensure maximum vigilance.

Now began the period of exhausting maneuvers, with the people forever in and out of the rigging. At times, the *Zélée*'s topmasts could be clearly seen, less than two ship's lengths distant, against a ridiculous blue sky, while vapor shrouded the rest of the ship. More eerie still was when both ships were fogbound in close company. Voices could be clearly heard on the deck of the *Zélée*, while the ship itself remained invisible. In the event of separation, the two ships were to rendezvous at Clarence

Island at the rim of the Antarctic circle. If that meeting failed, Valparaiso. Still no sign, and the surviving corvette was to return to France. Neither captain, under any circumstances, was to launch a southwards search for the other.

By the time they reached 62° south, still hundreds of miles north of Weddell's record mark, the company of ice mountains had become a thing of routine—a break in the monotony. Some they passed breathtakingly close, less than a ship's length, which gave them the opportunity for further study. As sunlight passed through the crystalline layers of an ice floe, the longer wavelengths attenuated rapidly into bursts of colors. Low, where the waves broke against them, the icebergs reflected an amethyst light, while higher up the whiteness of fresh snow was broken by long blue fissures, drawn at haphazard angles to the sea. The ice islands must have capsized at intervals on their northward journey, now horizontal, now vertical.

Morale was as high as it would ever be during the middle week of January. A pretty breeze from the southwest propelled them at six knots. They seemed, for now, to have outrun both the bad weather and the ice. The sea was calm, and they could make out ice islands fifteen miles distant. The commander's prediction of 82° south was now considered pessimistic by officers and crew. The prize for beating Weddell stood within their grasp. For the officers, men of means, glory was the main lure, but for the foremast hands the windfall held out the promise of a new life beyond the prison of the sea: a home, even a wife, and the turn to a respectable trade. But would the old man take them far enough to make their fortunes?

At three in the morning of the day D'Urville expected to pass 65° south, he was awakened by the lookout's incredulous cry. He was on deck in an instant. An immense pack of ice stretched uninterrupted from the north-east to the south-west horizon. The officers were already hurrying into the rigging.

From the mast, a remarkable sight. Instead of the sea, a vast undulating field of snow stretched before them, sprinkled with blue crystals. It had no visible end, nor was any passage visible through it. A jagged jumble of ice ten feet high, dense and impenetrable, dotted here and there with somnolent seals. An entire continent of ice. D'Urville, like James Cook seventy years before him, was appalled at the sight. In an instant, his illusions were shattered. There would be no 82° this year, no South Pole discovery. He felt a wave of disgust for James Weddell. The Englishman had either had amazing luck, or had fabricated everything.

Even in the midst of his cratering personal ambitions, D'Urville's mind focused, with piercing clarity, on what the appearance of the ice pack meant to his ships and his men. He worked through the political calculations in an instant. It would not be enough simply to have seen the ice pack and turned around as Cook had done. A voyage to 80° south and whatever lay beyond would have involved incredible dangers. But here, short of 65° south, he must risk even more to prove—first to his officers, then to a skeptical public back home—that he, Dumont D'Urville, had done all that was humanly possible to reach the pole. Looking out across the endless pack, he saw how far he must risk one hundred sixty lives just to save his name.

That first night in the Weddell sea ice, the moon rose like red fire behind the clouds, bathing the iceberg-littered sea with colors none of them had experienced or had words to describe. On the southern horizon, the ice pack appeared as a white line against the blue sea. Both sea and sky were calm, the only sound the *cri-cri* of penguins perched on the icebergs in their thousands, the occasional jet spray of a whale, and the incessant dull lash of the sea. It was a scene of primitive sadness, as if the world had abandoned them to a time before the Creation. Whatever upheavals of Earth had created this place, no memory of them existed in the minds of men.

The next day brought the worst possible turn of events: a whipping gale. Monster waves crashed in freezing sheets across the deck. More icebergs closed around them, while D'Urville was beset by familiar enemies: a headache like a piston hammer, nausea, and uncontrollable shivering. Several men had already been ordered below, fingers frostbitten and bloodied. From his little cabin erected on the poop deck, D'Urville ordered the sails trimmed as they rode out the storm through the long night.

The threat of instant destruction against the ice cliffs kept them alert even through their exhaustion. Thick snow swirled about the officers on watch, plastering their bearded faces. They could not see more than a ship's length beyond the rail. The wind rose first to a howl, then to a shriek so loud not even their firing a cannon could be heard through the muffle of snow.

During the brief night came sounds D'Urville had never heard in a lifetime at sea. The first was a deep boom, as if someone had struck a giant drum reverberating through the ship. The shock threw him onto the floor of his cabin. It was the sound of a ship helplessly beating itself against a rocky shore. Then came a cacophony beyond all description: a prolonged tearing noise as if some great sea monster were ripping the hull of the ship to pieces, plank by plank. An hour earlier, they had sat comfortably in open water. Now, the *Astrolabe* was beset. The ice had come to them, by stealth. The first blow signaled the sickening collision of the ship's hull with the ice, while the longer, scraping sound was her being dragged between ice floes. The pressure shook the corvette to the core, squeezing her for breath.

Was this deadly pack merely a belt of ice surrounding a warmer, open sea, as the optimists had dreamed of? Or was it landless, solid ice all the way to the pole, as in the Arctic? If the latter were true, to have met with its rim at 65° south implied a truly mind-boggling volume of floating ice that dwarfed anything found in the north. Which raised the question of why conditions should be so different in each hemisphere, and how the sea and subzero air, combined with whatever other natural agency, could maintain such a phenomenon. Was the ice delivered from some great hollow tunnel in the planet, as the cranks in America believed?

The only rational explanation left—shared by D'Urville and his officers—was that *there must be significant land to the south*. Not merely islands, as had been sighted by whalers like Weddell, but a great mountainous continent in the style of Patagonia, with soaring snowcapped mountain peaks and glacial ravines that sloughed their excess ice into the sea. If this were true, then they were within tantalizing reach of a truly monumental discovery.

An opening appeared in the ice, and D'Urville shouted the order to sail through at speed. It was the most reckless command of his career, but he gave it as he had a thousand others, automatically and without emotion. The Frenchmen soon found themselves in a sort of small, misty cove inside the pack, surrounded by walls of ice. The ever-obedient *Zélée* followed. The officers were jubilant. They had crossed the ice to Antarctica! A detachment from the *Zélée* arrived on an ice floe to drink a punchbowl of celebration. But D'Urville knew their "route" through the pack was nothing of the sort. It was a mere chance opening into which they had sailed—a channel that would now, according to the sinister nature of Antarctic sea ice, close behind them. They were trapped.

That night, the dismal music resumed aboard the *Astrolabe*. Blow after shattering blow staggered the ship until her commander was convinced she could not survive another hour.

His conscience plagued him. He feared his own death, in the common way, but his real anguish centered on the two ships' crews in his care. He had brought them to this miserable place only to abandon them to the drawn-out conclusion to their existence. He thought of the smile that news of their loss would bring to the vile face of François Arago.

The *Astrolabe* and *Zélée* spent the following two days turning in circles inside a small lagoon within the pack, their temporary sanctuary. Working the ship in this way involved repeated alterations of sail that D'Urville thought must break his men. On entering the ice, the air temperature had plummeted. A vaporous mist turned to icicles on their faces. The frozen ropes bit their hands red and raw. When the wind sprang up, conditions on deck were unbearable. The men sought respite in all corners below deck, rubbing their numb cheeks, and lighting pipes extinguished by the wind. D'Urville felt a constant glacial chill through his body, and his teeth chattered so he could hardly speak. Hypothermia at last forced him below, where he fell down and couldn't get up.

D'Urville sent one of his strongest hands, a sailmaker named Rougier, to explore northward on the ice pack. Rougier returned with grim news. He had trekked two miles at least. As he approached the rim, still far in the distance, the ice grew larger and more dangerous. Beyond the distant open water appeared to be yet another barrier of ice. D'Urville climbed into the rigging, taking his best spyglass with him. Even at the dizzying height of the mainmast, he could not see beyond the ice field. It enclosed them at every point, stretching to the grey horizon. Only to the north could he make out a tiny ribbon of darkish blue that might suggest open sea. Or it could be an illusion.

He set a course for north-northwest, knowing full well that filling his sails would involve continuous collisions with the ice.

Fig. 4.2. Louis le Breton's dramatic depiction (1841) of the French corvettes at the mercy of Antarctic ice was designed to revive public interest in exploration in the waning days of the Age of Sail.—Musée des Beaux-Arts de Quimper.

But simply heaving to was nothing better than surrender. Within minutes, a thick fog enveloped them, then swallowed up the *Zélée*. The anxiety was unbearable. Even if they were to approach the northern rim of the pack, this would bring them within range of the giant swell from the open sea. Then any change of wind would run them back against the brick-hard ice and they would spend their final hours watching their ships disintegrate piece by piece beneath them.

As a spectacle, their situation was truly unique. D'Urville struggled to find words to describe it in his journal. Hour after hour, the immense pack rose and fell with the swell of the sea, buffeting the ships. It was as if they were surrounded by volcanic islands of snow shaken by eruptions, one every few seconds. Meanwhile, the dazzling specular monotony of the pack began slowly driving them crazy. The ice took on taunting shapes: a house, a familiar streetscape, a cathedral in Paris. At these high latitudes, in midsummer, daylight persisted until eleven. A bare two hours later, a gray crepuscular light returned. As sailors accustomed to four-hour watches, the *Astrolabe*'s crew slept at will with no regard to the sun. Still, the never-ending day and remorseless glare began to really distress them. They complained of shooting lights in their vision: "ice blindness." D'Urville took to wearing sunglasses on deck.

Numerous sightings of land had all proven false. Then one morning, around nine o'clock, D'Urville, who had the best eyes of anyone on board, saw the clear outline of land on the southern horizon: a rugged, mountainous terra firma terminated at each coast by a long, level point. He called Dumoulin to make a record of their discovery. Word spread quickly through both ships. If they could not beat Weddell's southing, they could now lay claim to something far greater: discovery of an Antarctic continent, beating out the Americans and British. Dumoulin sketched busily for an hour, then slowly lifted his pen. The land, just now so clear, was changing shape before their eyes, deforming itself into mirage. Mountain peaks turned to pillows, ridges to ripples, as their fortune faded from view into the blue-white sky.

Feelings of powerlessness began to overwhelm them. The pack was silent and gloomy. The human senses had no purchase here: no vegetation, no smell, only the same unbroken whiteness as far as the eye could see. If they were to come to grief and be marooned, there would be no consolation. Perhaps a poet would have taken pleasure in the view, but for the officers and men of the imprisoned *Astrolabe* and *Zélée*, the scrawniest olive grove in Provence would have been more delightful to the eye than this awful sublimity day after day. They had sailed beyond all oceangoing rules, beyond calculation of risk. It was a world of pure, inhuman chance, where the odds of survival grew longer each day.

At night, more loud booms, like detonating explosives, woke D'Urville from his uneasy sleep, and he knew that some collapse was underway in the kingdom of ice. The ship's side, where it had scraped hard against the pack, left a long streak of brown copper on the wall of the ice. Like a drunken man in an alleyway, the *Astrolabe* staggered from side to side in a crazy, slewing motion, recoiling from each blow to its hull. At any moment, a gust of wind from the wrong quarter, joined with the ocean swell and an ice floe cruising just so, would smash the ship to pieces. As long as the wind persisted from the north, their position was hopeless. The southern summer had loosened its fragile hold. The melting ice was refreezing before their very eyes, sealing the ships within it.

The cruelest night of their entire ordeal began when D'Urville was awakened from a deep sleep by Lieutenant Marescot, full of excitement. The commander rushed on deck. To his astonishment, the sea had cleared a channel for them. Their saving wind was drifting from the east-southeast, carrying them to open water. He immediately gave the order to make full sail, but Marescot and Roquemaurel, officers of the watch, looked at him gravely. The light was fading in a heavy snow. He was asking them to sail blind. He rescinded the order and went below again. The next to wake him was not an officer but only the poor helmsman. "How's the ice?" D'Urville demanded. Reluctantly, the man replied, "As it was last evening." D'Urville, enraged, dared him to repeat himself. As the news sunk in, he drew the information from his subconscious mental log that the ship had not moved in the last half hour. On deck, he saw that their situation had reverted. The *Astrolabe* was besieged by ice as tightly as ever.

With the ships stuck fast, D'Urville resorted to desperate measures. He ordered the forward anchors to be fixed a hundred yards distant on the ice while the men worked at the capstan, expending their last ounce of strength to turn and coil the ropes inch by inch. Four hours pushing the spokes of the capstan wheel yielded a half mile of progress. Then suddenly, the ship surged into empty water, sending them sprawling across the deck. The men at the anchors, blinded by the snow, raced back across the ice to rejoin the emancipated *Astrolabe*, for there was no turning back. One man, called Aude, was given up for lost, only to be hauled in at the last possible moment, frozen and half dead. He had been one of their strongest men but never recovered.

The ice was not a smooth and pristine plateau but heaped up, confused, and dangerous—a primal chaos. Waves lashed the fleeing ships, surging across the bows in a freezing waterfall. At one such blow, only Lieutenant Roquemaurel's quick thinking prevented D'Urville from pitching headfirst off the poop deck: he grabbed the sleeve of his commander's coat and hung on. The hull resounded with the blows of constant collision with the ice, while all eyes lifted toward the masts, which bent sickeningly low before righting themselves in a clatter of rope and canvas. The loss of a single mast would be fatal here, hundreds of miles from any shipyard or forest. And winter in the ice would bury them all.

Just when they were beyond exhaustion, an open sea appeared to the north, two miles distant. D'Urville shouted for everyone to get back on board. Soon they were among the thinner, shifting ice, pitching and tossing while frigid waves drowned them twice a minute. All at once, they shot like an arrow through the sheer rim where the ice pack met the open water. They felt the real sea swell beneath them. The men shouted their relief like overtired children. The monthlong burden of disaster lifted instantly from D'Urville's shoulders. He felt like a deposed king restored unexpectedly to his throne, back in charge of his little kingdom's destiny.

They turned for a last look at the ice pack retreating behind them. Through a veil of sky, they could see its deceptive blink at the horizon. Then came the caulker's report. The *Astrolabe* had not leaked a single drop of water during their drawn-out death match with the pack. No one could believe it. They had passed beyond the forbidden boundary of the south polar ice—a natural human limit where neither James Cook nor the Patagonian icemen had ever ventured—and escaped with their lives.

But even in the raw euphoria of survival, D'Urville felt a sinking unease. With every degree north they sailed, the story of the *Astrolabe* and *Zélée* in the ice pack would tell more like a retreat than a triumph. By the time the tale reached Paris, he guessed, they would be calling it a scandal for D'Urville to have reached no farther than 65° south, hundreds of miles short of Weddell, with not even his noble corpse to show for the effort. Unlike the more fortunate Ernest Shackleton of 1912, a glorious failure on the Weddell Sea ice was not an option for Dumont D'Urville. Though every fiber of his body and mind rebelled at the idea, he knew he had no choice but to make plans for a second assault on the pole.

Interlude: Ice Station Weddell

The work of the Ocean Drilling Program in the Weddell Sea has helped establish a timeline for Antarctica's long-ago separation from Tierra del Fuego and the South American continent, creating the stormy strait across which Dumont D'Urville sailed his corvettes in January 1838.

Fifty-two million years ago—during the so-called Eocene Climatic Optimum, when thick forests populated the poles—a land bridge connected South America and what is now the Antarctic Peninsula, allowing the free exchange of long-vanished plants, birds, and animals. It was the last hurrah of old Gondwana before Antarctica's farewell. America drifted north. Continental crust subsided, flooding the transit zone. By the time of the Eocene-Oligocene Transition, a mile-deep waterway had opened: the Drake Passage. The Pacific and Atlantic Oceans were now connected south of Cape Horn, and a new ocean encircled Antarctica, isolating it in deep freeze. Today, the massive energies of the globe-girdling Southern Ocean squeeze through the narrow, five-hundred-mile Drake Passage, creating the most powerful current in the world and monstrous seas. In high summer, the Drake Passage becomes a fast-flowing exit ramp for icebergs calved from the giant glaciers of West Antarctica and its northern peninsula.

Looking at the map, the Antarctic Peninsula extends, like a finger, beyond the Antarctic Circle itself toward Tierra del Fuego, to which it was once joined. If we imagine the glaciated East Antarctic coast as the thumb, then in the hollow dip between finger and thumb, south of the Drake Passage, lies the Weddell Sea, a spectacular hub for Antarctic wildlife, sea ice production, and oceanic energies. The fact that the unknown icy waters into which the *Astrolabe* and *Zélée* sailed in 1838 are today labeled the "Weddell Sea" and not the "D'Urville Sea" tells the story of the French vessels' thwarted southern campaign that summer.

The Weddell Sea is nevertheless rich in polar exploration lore. In 1912, Ernest Shackleton's ship *Endurance* became stranded in the Weddell ice pack and sank at 68° south, a disaster that set in train one of the greatest of all polar survival stories. Eight decades later, in the last days of the Cold War, American and Russian scientists joined forces to reenact D'Urville's ordeal, and the final, rudderless journey of the *Endurance*, by erecting a temporary station on the Weddell ice pack. Ice Station Weddell, led by Arnold Gordon of the Lamont-Doherty Earth Observatory at Columbia University, ranks as one of the most unlikely, and daring, scientific installations of the late twentieth century.

Because of its perennial ice, the western Weddell Sea, adjacent to the Antarctic Peninsula, had always been inaccessible to research vessels. With Ice Station Weddell, Gordon's objective was to collect data on this forbidding mystery ocean. Pitched on undulating ice no more than a meter thick, buffeted by gale force winds shearing off the Antarctic continent, and in temperatures that dropped to −35°C, the ice station scientists traveled seven hundred fifty kilometers over the course of four months in 1992. As they drifted along at the whim of the ice pack along the western 53rd meridian—a course that crisscrossed that of D'Urville's *Astrolabe* and followed a route strikingly similar to that of Shackleton's *Endurance*—Gordon and his colleagues gathered vital original data on the behavior of Antarctic temperature, currents, winds, and sea ice. The Weddell Sea, it turns out, is a scene of ocean mass modification unique on Earth.

The annual expansion and retreat of sea ice in the Southern Ocean—by as much as two thousand five hundred kilometers at certain points—is spectacular physical evidence of our glacial climate system and its seasonal rhythms. At its winter maximum, the Weddell Sea ice pack extends across eight million square kilometers of ocean. In the peak summer conditions into which D'Urville led the *Astrolabe* and *Zélée*, that seasonal ice was "reduced" to a rump perennial pack twice the size of France.

In the western, coastal zone of the Weddell Sea, supercooled open waters by the glacial continental shelf function as a factory for sea ice production. The ice begins as microscopic crystals suspended in a water column perpetually agitated by strong offshore winds. In time, a thin slush forms on the surface, insulating the crystals from the harassing wind and allowing ice to grow in the dark and quiet subsurface. Beginning as a thin, granular pancake, the newborn ice concentrates salt in the water column, driving cold, dense water through a warm middle layer down to the abyssal depths at the rate of a million cubic meters per second. This Antarctic Bottom Water—70 percent of which originates in the Weddell Sea—then curls irrepressibly northward into the Atlantic and Pacific basins, chilling the bottom kilometer layer of the world's oceans. The thermohaline pump in turn delivers upwelling warmth to the tropics and midlatitudes, maintaining our modern climatic zones.

The most remarkable discovery of the 1992 Ice Station Weddell project was that this global ocean conveyor belt—a physical phenomenon of planet-defining dimensions—originates in small-scale processes along the coastal shelf of the Weddell Sea. (That the Russian scientists were able to contribute to this research achievement while the Soviet Union was disintegrating around them, like so much summer sea ice, was no less remarkable.) Freezing southerly winds off the Antarctic continent power a cyclonic ocean current called the Weddell Gyre, motoring offshore at the rate of twenty-eight million cubic meters of water per second. In a kind of ice accordion effect, these winds maintain the frigid temperatures necessary for permanent ice cover while also driving the

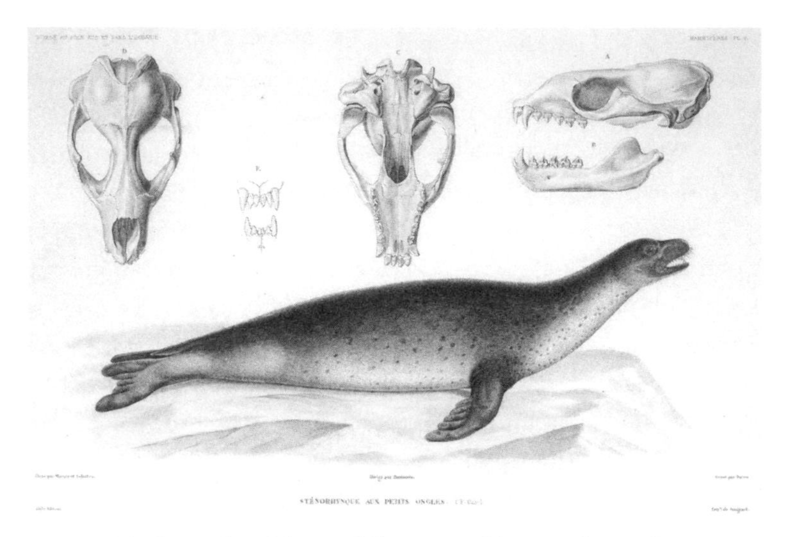

Fig. 4.3. The leopard seal, beautifully captured by French expedition artists, may be found all across the Antarctic Ocean, a natural resident of the ice pack. The French explorers could also boast the first description of the heretofore unknown crabeater seal (*Lobodon carcinophagus*). Seal populations were decimated across the sub-Antarctic islands in the 1820s and '30s.—Dumont D'Urville, *Voyage au Pole Sud et dans l'Océanie sur les Corvettes L'Astrolabe et La Zelée . . . Zoologie* (1842–53). Biodiversity Heritage Library/Smithsonian Libraries.

pack northward along the gyre to open coastal waters for the creation of new ice. Between its formation and final melting in warmer northern waters, an ice floe may travel more than a thousand kilometers.

As the engine room of the oceans' deep cold, the Weddell ice pack reflects 90 percent of the sun's rays back to the atmosphere. To the homesick sailor's eye, it is a theater of illusions. Sun-drenched icebergs look like churches, ice floes like a village green. This weird icescape, in which D'Urville's polar nightmares were fully realized, depends on the perpetual freezing and unfreezing of open water, as deep warmer currents advect to the surface, come into contact with the freezing air, and release their vaporous heat to the atmosphere. For the *Astrolabe* and *Zélée*, the high-energy congress of ice, ocean, light, and fog in the Weddell Sea made for a baffling, infinitely treacherous environment. Ice-free lagoons—or polynyas—promised all-too-temporary sanctuary for the ships, while open water tracks between ice floes held out fleeting avenues of escape. During their life-and-death struggle with the Weddell ice pack in the summer of 1838, the constant fluctuations of open water amid the ice—and the exhausting cycle of hope and disappointment it entrained—kept the French commander and his crews on a knife's edge of sanity.

‡ 5 ‡

The Voyage of the *Flying Fish*

Some six months after the French expedition's escape from the Weddell ice pack—in November 1838—the man who would botch the American Antarctic campaign, and thereby cast the US Exploring Expedition into a century-long oblivion, peered into the stinky darkness of the *Peacock*'s hold. Lantern held high, he listened. Water was flooding in from all directions. It was as if every seam in the ship had rotted and was leaking. Not only the decks, but her coamings and her very stem—all unsealed to the sea. On deck, the mildest of breezes blew, but to judge by the amount of water *Peacock* shipped, she might have been riding out a gale. He had just come from his cabin, which was flooded to knee height, and where he had woken to the sight of his rug floating away.

Lieutenant William Hudson—commander of the *Peacock*—pushed his way into the storeroom, where a waterfall was gushing down the apron onto the floor. He quickly went aft to the pump well. The men would be at the pumps all the way to Rio to keep the *Peacock* dry and afloat. Hudson was not entirely surprised. He knew the ship had not been ready to sail out of Norfolk. Wilkes had put the squadron to sea in August 1838 only because he had promised the secretary of the navy that he would, to relieve Washington and the nation from the embarrassment of the "Deplorable Expedition." He and Wilkes understood the *Peacock* would require a full refit in Rio, at great expense of time and money.

What Hudson was not prepared for—what truly staggered him—was the state of the pumps. He climbed down into the well and made his inspection in the shimmering pool of the lantern light. The iron bands that secured the pumps were entirely rusted through—useless. One set of bands had fallen off and lay in pieces on the floor. This was worse than a hurried departure out of Norfolk, or negligence at the Navy Yard. For a ship destined for Antarctica, it was better described as mass murder.

Once assembled in the magnificent surrounds of the bay at Rio, Wilkes and Hudson conducted a thorough reassessment of the American squadron handed to them to conquer the South Pole and circumnavigate the globe. The flagship *Vincennes* dominated the group, a fine-sailing sloop and the pride of the Navy: broad-beamed but fast. She shipped one hundred ninety men. At a third the size, the brig *Porpoise* outsailed the *Vincennes* in a light wind. Wilkes had a sentimental attachment to *Porpoise*—he had recently commanded her on a survey of the Georgia coastline—and added her to the squadron after the Catesby Jones fiasco. To Wilkes's chagrin, the one ship commissioned by his predecessor that he had not rejected—the storeship *Relief*—had proved a complete failure on their Atlantic crossing. Beautifully appointed, she sailed like an overpriced slug, setting a new record of a hundred days for the slowest passage from Norfolk to Rio.

But it was not these ships that excited the attention of the expedition's officers. Instead, they all looked on the little schooners, *Sea Gull* and *Flying Fish*, with almost indecent lust. Command of the schooners—tenders to the *Vincennes*—lay at the disposition of Commander Wilkes and would represent, for any of the dozens of ambitious young officers attached to the expedition, the first longed-for taste of command. At a mere seventy feet in length each, with crews of fifteen, the schooners did justice to their namesakes, darting across the Atlantic as in their natural element. More than that, these delightful craft were American in design, adding to the crews' perception of their beauty.

Wilkes was painfully sensitive to the issue of seniority, and so were his officers. He had already caused genuine anger in the ranks for his appointment of two mere midshipmen to the command of the *Sea Gull* and *Flying Fish* on their outward passage. Now in Rio, where he had finally revealed his plans to stage a late assault on the South Pole, the question turned to whom he would choose for command of the schooners on this mission, with its rich promise of death or glory. For the two officers he chose, it would represent a career-making opportunity. A first command signified so much more, after all, if it was also in danger of being the last.

It was apparent to all that Commander Wilkes nourished a prejudice against the officers he had inherited from Catesby Jones. The more competent that officer, the greater the animosity. First, Wilkes had taken a strong dislike to his second-in-command on the *Vincennes*, Lieutenant Craven, who managed the ship with smooth professionalism. Craven had been relieved of his station and now languished in port aboard the *Relief*.

When it came to distributing the great prizes of the southern mission—the *Sea Gull* and *Flying Fish*—the same prejudice prevailed. Wilkes understood the necessity of replacing the midshipmen but saw fit to ignore the rights of seniority among his eager lieutenants. Two junior lieutenants, William Walker and Robert Johnson, became the envy of the fleet, while officers senior to them were passed over. Walker and Johnson had in common that they had joined the expedition after Wilkes's own appointment. A Jones ally named Lieutenant Clairborne outlined his grievances in a formal letter. Wilkes responded by dismissing him from the expedition. He was advised to catch the next boat home. When another senior lieutenant, named Lee, complained, Wilkes sent him packing too. The rest of the disgruntled officer corps was stunned into submission by Wilkes's actions, just as he intended.

Meanwhile, Brazilian slaves recaulked the *Peacock* in the harbor at Rio, at great expense to the United States Government. William Hudson had served on the West Africa squadron and so knew the slave trade. There was no depth of human degradation he had not seen. But for many of his young, sheltered officers, Rio made for a painful awakening. Chain gangs filled the streets, while slaves past their usefulness—old and sick—wandered as beggars. Nothing compared, however, to the sight onboard the slave ship brought to port by an English merchantman. African captives were crowded on the open deck under a burning sun in their own excrement. Many lay dead, while the rest were gaunt and blank-eyed from starvation. Below deck—what were the conditions there?—the rumor had spread that the human cargo was to be sold to cannibals. Twelve captives had killed themselves. To the Yankee abolitionists aboard, the prospect of a pure southern horizon, empty of humanity, became morally inviting.

From their home base in Orange Bay, in Tierra del Fuego, Wilkes set about reorganizing the fleet for the polar voyage with a brutal logic worthy of a slavetrader. The scientists were packed aboard the *Relief*, while the "disgraced" Lieutenant Craven took command of the *Vincennes*. Neither ship was to share in Antarctic glory. Wilkes himself took temporary command of the *Porpoise*. She and the *Sea Gull* were to retrace Weddell's legendary course beyond the Shetland Islands to the southeast, while the *Peacock* and *Flying Fish* would sail westward to longitude 105° and attempt to beat James Cook's record southing in those waters.

Wilkes's orders to the commanders of the *Peacock* and *Flying Fish* stipulated that they should keep close company and guard at all costs from being separated. But orders mean little in the teeth of a gale, which the two ships encountered west of the rocky islands of Diego Ramirez. Hudson had ordered the regular firing of the cannon and had blue lights lit on the masthead, all means of keeping contact with young Walker aboard the *Flying Fish*. But after forty hours of riding out the storm, another fourteen waiting for the schooner to bear up, and lookouts straining at all horizons—nothing. Within forty-eight hours of departing Orange Bay, the *Peacock* had lost her consort.

Meanwhile, from the cramped deck of the *Flying Fish*—still drying out from the storm—William Walker was sure he could make out the *Peacock* hull up to the north, heading west. He and Commander Hudson had settled on four rendezvous points in the event of their separation. In the vast landless expanses of the Southern Ocean, this meant locating a precise, arbitrary point in the heaving open sea and laying to. Beating up into the wind, across the giant waves, the *Flying Fish* began leaking badly. Standing upright on deck became impossible, while below deck resembled a swimming pool. Books, clothes, and furniture sloshed from side to side. In Walker's tiny cabin, the freezing water rose above his waist.

On the third night of a continuous gale, the schooner's jib sail split right down the middle in the howling wind. Battening down at the first rendezvous point, a rogue wave crushed the two lifeboats to pieces and washed the binnacle overboard. The helmsman and the lookout crawled below with bleeding limbs. Meanwhile, the Antarctic wildlife seemed to find en-

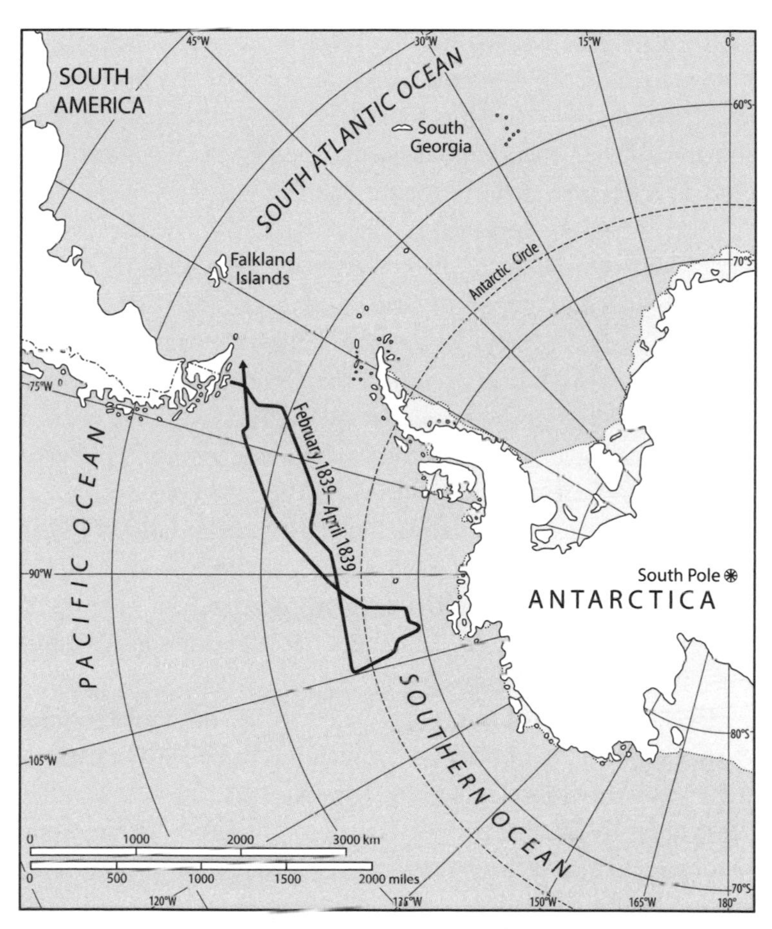

Fig. 5.1. The record-setting voyage of the *Flying Fish*, March 1839.

tertainment in their plight. A whale rubbed his vast sides against the hull, while an albatross hovered serenely above the deck.

No sign yet of the *Peacock*. When the shrieking wind at last fell an octave, the crew discovered a leak in the bread room. All hands gathered to shift the stores aft. The *Flying Fish* was now in deep trouble, so Walker decided to defy his orders and ignore the remaining points of rendezvous for the time being, and instead allow the ship to sail as the winds blew. Farther southward then they went, accompanied by flocks of birds including a polar sheathbill, whiter than snow.

Fleeing south, conditions only worsened for the *Flying Fish* and her frostbitten, battered crew. They were all day at the pumps as mountainous seas swept the decks. In the brief stillness of the troughs, each feared the next wave might be the schooner's last. Every seam of the *Flying Fish* leaked, every man was soaked to the bone, every bed and trunk afloat. On deck, they staggered around with their feet wrapped in blankets, to stave off frostbite, while the sideways sleet froze on their clothes, encasing them in ice. Soon, fully half Walker's crew—five men—were disabled by fractured ribs, body bruises, and hands in bloodied shreds from hauling on icy ropes. Still, no one complained. The sheer habit of duty prevailed—for now.

Two sunny days at last, with clear sailing, brought the men within sight of their prize. Two enormous icebergs, like a south polar gateway, loomed into view. They reached 105° west, Cook's longitude, in a sea studded with ice. This knowledge alone made them briefly forget their physical suffering, and they strained their gaze to the mysterious south. Bergs a hundred feet high were covered with penguins, and whales filled the sea, so that the crew occasionally needed boat hooks to push them off.

It was no longer the open ocean, nor like any seas they had ever known. Fog was perpetual, an alternation of mist and snow. Waves broke against the sheer ice in a thunderous slap, or sent a human-like roar from the dark interior of a cavern of ice, sculptured like some gloomy scenery for the stage. During the brief, shadowy nights, the officer on watch kept lookout at the bow, trusting his ears to guide the schooner between the invisible cliffs of ice.

For three days, the vessel lingered in sight of land—or the appearance of land. Ice drifted alongside streaked with dirt. The water grew thick and dark. The siren lure of an undiscovered coast drew Walker farther into the ice labyrinth, farther south than his schooner belonged until, in the imperceptible passage from one hour to the next, he became aware that the thousands of independent ice islands through which they had sailed were now closing ranks into a single, impassable mass. Then this, too, broke up without warning, and it seemed fate was smiling upon them when a clear sea opened to the south. Soon they had passed the southern mark set by the Frenchman D'Entrecasteaux, then the Russian expedition leader Bellingshausen in 1820. Cook's mark was next, not a day's sail farther, with a fair wind from the north to assist them. Their sails were in rags, but Walker set every last one.

The next day saw a turn for the worse. A fresh gale forced Walker to heave to and, during the grainy night, they heard a deep rumbling to the south. All of a sudden, the fog lifted to reveal the *Flying Fish* beset by ice. The men staring at the pack were soon covered in hoarfrost, like specters. The oppressive silence produced hallucinatory sensations. Words died on their lips, so they could not be heard by the man standing next to them. The heaped-up ice, played upon by blue and green light, took on the appearance of land, even a familiar-looking town complete with church and steeple. Then the light lengthened to the horizon, unveiling a monotonous white world without end. Mountains of ice blended with the whiteness of the sky. There was nowhere for the eye to rest, nothing for the machinery of mind to grasp.

Fig. 5.2. *The Flying Fish in a Gale* (1839), by the American expedition artist Alfred Thomas Agate. Navy Art Collection.

Walker realized, too late, that he had allowed the schooner to drift inside the main barrier of ice, even as her northward exit closed silently behind her. Large oblong floes of ice rotated around the hull. He understood now the meaning of the clicking-crackling sound in the gloom. The sea was freezing over, seizing *Flying Fish* in its grip. The snow that fell on the water no longer melted but froze up and covered the waves. For the first time, he felt the eyes of the crew upon him. They had endured everything until now without question. But this was a death zone, where duty no longer applied. He had taken them beyond the range of his command.

If youth played any role in Walker's risk-taking course into the ice, it certainly fired his headlong effort to escape. A lucky breeze sprung up, and he ordered the *Fish* to be driven, bow rearing, onto the ice. She careened wildly from side to side as she lurched northward, twisting like a corkscrew, smashing her hull on the ice until splinters flew. The masts bent and buckled like reeds at each concussion. When the very bolts began to fly loose from the timbers, the carpenter ran forward to tell the captain to stop before the ship fell apart beneath them. But Walker didn't listen.

Waiting anxiously several degrees to the north, Commander Hudson of the *Peacock* thanked God Almighty at the sight of *Flying Fish*, beating her way through the broken ice. Over dinner in the captain's cabin, they all marveled to hear Lieutenant Walker's story of how far into the pack the little schooner had ventured, beyond Bellingshausen, then to Cook's southerly mark. How her crew had been trapped there, and how their thoughts had turned to home and their loved ones. Then followed Walker's mad dash, which saved them.

Back at Orange Harbor in Tierra del Fuego, where the entire squadron were now gathered excepting the sluggard *Relief*, it quickly became apparent that Walker and his *Flying Fish* were the heroes of the hour. A tender of ninety tons had outdone the glamorous *Porpoise* and *Peacock*. Wilkes in the *Porpoise* had lasted a mere four days on his southern course before meeting the ice pack and turning around, with the disappointed *Sea Gull* in tow.

Citing the lateness of the season, Wilkes now officially abandoned the Antarctic mission for the year, leaving his cynical officers to wonder why they had not embarked on it earlier. Because he kept all details of their expedition to himself, Wilkes did not tell them that his orders were to make a second attempt on the South Pole in the summer of 1840, this time from the east, from Sydney, Australia. Lacking this information, the officers were left to draw their own conclusions. Sensing their chance at glory slipping from their grasp, they fixed on Wilkes a penetrating hatred. Now that he had aborted their southern mission, the officers fully expected him to return the command of the schooners to lowly midshipmen. This he did, without explanation.

On the northward leg from Orange Harbor to longed-for shore leave and rest at Valparaiso, the squadron met with a pounding gale. Hurrying to the sanctuary of a cove, they found themselves without the *Sea Gull*. As weeks passed, fear grew to certainty that the *Sea Gull*, an inexperienced midshipman at the helm, had been lost with all hands. Wilkes expressed no remorse. Nor did he hesitate to sell her sister schooner *Flying Fish* when the opportunity arose later at Singapore. He considered the vessel surplus to his needs. So it came to be that the valiant *Flying Fish*—whose death-defying southern campaign along the Antarctic Peninsula in March 1839 must rate among the greatest American feats of sail—passed into foreign hands without a second thought.

Interlude: The Stormy Petrel

The Southern Ocean winds—which may be refreshing, bracing, or life-threatening depending on their strength and persistence—are an expression of difference in atmospheric pressure across space. Air pressure, in turn, depends on a declining gradient of temperature and humidity from subtropical latitudes to the poles. Without this gradient—without our glaciated poles—Earth's climate would be unrecognizable. Sailing north to south across the great ocean that surrounds Antarctica, balmy temperatures and smooth temperate seas give way to a cyclonic maelstrom—a ships' graveyard. The closer one sails to Antarctica, the sharper the temperature difference between the continental dome of ice and the relative warmth of the surrounding maritime air, generating storms of epic violence.

In the southern winter of 1838, the French polar expedition under Dumont D'Urville, which had led the tri-nation race into Antarctic waters the summer prior, escaped westward from Chilean ports into the haven of the South Pacific. On August 1, the *Astrolabe* and *Zélée* anchored alongshore the tropical island of Mangareva, glad to be done with their two-month voyage across virtually landless ocean. The French didn't know it, but they had reverse-traced the approximate meridional course of the brutal storm that would, nine months later, barrel thousands of miles across the southern Pacific Ocean to sink the rival Americans' schooner *Sea Gull.*

Storm track data were sparse across the Southern Ocean until the International Geophysical Year (IGY, 1956–58) established twenty new stations in Antarctica and spurred observations of the southern upper atmosphere. In the aftermath of the IGY, Hubert Lamb, pioneer of historical climatology, published the first description of Southern Hemisphere extratropical cyclones, which do not generate evenly or at random, but follow favored paths from the midlatitudes southeast toward the pole. One such track, Lamb concluded, has its genesis in the waters south of Tahiti and curls poleward to Tierra del Fuego and the Drake Passage before hooking north and dissipating in the Atlantic. In July 1957, during the IGY, no fewer than eighteen midlatitude storms were recorded passing from the west through the Drake Passage.

From Lamb's sketch, we can plausibly reconstruct the storm that doomed the *Sea Gull.* Sometime in mid-April 1839, northeast of New Zealand, a surface mass of warm equatorial air met with Antarctic currents tracking north: a so-called polar front system. The dense, cold currents undercut the warm air, producing first a wave, then a cyclonic spiral of low pressure. Within this steep temperature differential, an awesome reservoir of potential energy became available. Steered by the prevailing westerlies, and without any landmass to drain or deflect its power, the new cyclone set out to fulfill its destiny: to redistribute hemispheric volumes of water and energy from the northwest to southeast, across the watery expanse of the Pacific and Southern Oceans. The fact the crowded shipping lanes around Cape Horn lay directly in its path was irrelevant, meteorologically speaking.

As the massive storm system developed, cold air bands wrapped and sequestered the warm air aloft, maximizing its intensity. Squalls spun off the advancing front, spawning winds of hurricane force. A week after it first ruffled the placid Tahitian waters as a welcome breeze, the monster storm reached the Drake Passage south of Tierra del Fuego, where the most powerful ocean currents in the world joined forces with the atmospheric tumult to engineer a maniacal climax. Off Staten Island near Cape Horn, howling winds and waves overwhelmed the poor *Sea Gull*, which, though more than half the length of the flagship *Vincennes*, was barely an eighth her weight—slender, shallow, and exposed. The storm of April 29, 1839, likely ripped the foremast clean out of *Sea Gull*'s deck, smashed her timbers, and sank her. At least the end would have been quick for the fifteen crewmen. Her debutant captain, James Reid, son of the governor of Florida, left behind a young wife and a child he had never seen.

Sailors are proverbially suspicious and bestow names on their craft like wishful superpowers. Open ocean birds—gifted riders of the storm—are naturally popular choices. The nimble *Sea Gull*, it was hoped, would weather the South Atlantic gales with avian nonchalance. Likewise for the fleet's commander, whose moniker "The Stormy Petrel" was perhaps the politest term by which he was referred. The nickname expressed an implicit hope that, bad temper aside, Charles Wilkes might be granted the hardy powers of his seabird namesake and bring them all back to Virginia in one piece. By the same logic, however, the fate of the *Sea Gull*—arguably Wilkes's fault—did not augur well for the Stormy Petrel, or for the men under his command.

The Southern Ocean is synonymous with its remarkable indigenous birds. Reading Cook's famous *Voyages*, the English poet Coleridge was inspired to write a nightmare tale of sailing the vast Southern Ocean toward the ice, and did so through an unforgettable psychodrama of man and bird. His ancient mariner is "alone on a wide, wide sea" but for the albatross, which keeps him wary company. The bird's gratuitous killing at the sailor's hands sets in train a world of horrors. In the real-world Southern Ocean, not only the albatross but an airborne fleet of petrels, skuas, gulls, terns, cormorants, prions, and gannets escorts the human intruder toward the pole. A southern seabird first appears at the stern of the ship and lingers there, relishing the updraft. For variety's sake, it will sometimes glide across the bows, giving the observer on deck a view of its taut underbody in flight (perhaps it was this view that tempted Coleridge's mariner to murder). Even when a ship like the *Sea Gull*—and all who sail her—is fighting for survival in a deadly tempest, the same bird will continue to swoop and swoon about her pitching decks without regard for the turmoil below, as if to mock human helplessness. Southern Ocean gales hold no terrors for the birds, which, once fledged, never return to land except to mate.

In a landmark experiment in the late 1960s, to unlock the mysteries of avian flight, Colin Pennycuick filmed pigeons in a wind tunnel. His interest lay, in particular, in birds' remarkable resilience in gale conditions. Drawing on his knowledge of helicopter engineering, he imagined the bird as a circular disk—the limit of the circle described by the wingspan—creating, through participation with the wind, a downward resistance to maintain its gliding body aloft. The ratio of flight speed to bird power generated a celebrated U-curve that revealed the pigeon's capacity for minute in-flight adjustments. Birds, it turns out, have *two* optimal speeds: one for minimum energy expenditure, one for maximum range. In the real world beyond the wind tunnel, a bird might toggle from one speed to the other depending on its desire to forage, migrate, or nest, while the desired velocity itself varies, second by second, with the ever-changing currents of the atmosphere.

Publication of his breakthrough paper brought Pennycuick's opportunity to fill his ornithological bucket: to graduate from study of the common, lowly pigeon to observation of charismatic birds in exotic locations. The Southern Ocean seabirds, which spend their lives cruising the windiest seas on Earth, are the pure objects of bird aeronautics. Accordingly, the summer of 1979–80 found Pennycuick on the sub-Antarctic island of South Georgia—home of the most populated nesting grounds in the Southern Ocean. He came equipped for a three-month observation of the flying style of the legendary albatross and Antarctic petrels, members of the *Procellariiformes* bird group.

From a rocky outcrop on the northwest tip of the island, the birds appeared above the high coastal ridge, slope-soaring along the cliff face, then headed out to sea. They flew into or across the wind, not with it, tacking in a zigzag course along the crest of the waves. The drooped leading edge of the albatross wing enabled the giant birds to catch the updraft from each passing wave, maximizing lift with minimal fuel. Of all Antarctic seabirds, the albatross, in its physique and function, is the most dedicated to exploiting the energy potential of winds as they ruffle and whip the sea's ever-changing surface.

On the dissecting table, Pennycuick uncovered the engineering secret of the albatross. A fan-shaped tendon, stretched across the outer breast muscle from sternum to humerus, locked the wing to a horizontal plane. Evolution had determined that no bird should fight the Southern Ocean winds to maintain stability or, in gale-force winds, risk a bone-breaking calamity (think of the *Sea Gull's* shattered mast). Except in a rare perfect calm, the birds of the Antarctic Ocean don't bother to beat their wings. As each wave displaces stationary air upward, the albatross accelerates. At maximum speed, it banks suddenly—a breathtaking maneuver—converting excess kinetic energy to potential energy as it abandons one wave for the next, coolly sacrificing speed for position.

There could be no greater contrast with the glorious, aerodynamic ease of the albatross than the helter-skelter, almost frantic flapping of the Wilson's storm-petrel, for which Charles Wilkes was aptly nicknamed. The storm-petrel was the smallest of the procellariiform species included in Pennycuick's study. While 1 percent of the body mass of an albatross will power open ocean flight for nearly thirty hours, the swallow-like storm-petrel will burn as much body mass in two hours, zipping headlong between the waves.

But regardless of size, the seabirds of the Antarctic share perfect adaptive traits. Calculated across distance, their apparently erratic course represents the shortest route from nesting to feeding

Fig. 5.3. The snow petrel is the southernmost bird on Earth. It breeds exclusively in Antarctica and has been observed at the South Pole. The "petrel" name is inspired by Saint Peter, the birds seeming to walk (actually, run) on water during takeoff.—John Richardson, *The Zoology of the Voyage of the HMS Erebus and Terror* (London: E. W. Janson, 1844–75). Biodiversity Heritage Library/Woods Hole Library.

grounds. Most importantly, the wing engineering embeds millions of years' worth of ocean storm experience, a long-term species data-gathering effort that enables any individual albatross or petrel—observed, say, from a South Georgia station or the decks of a foundering explorer vessel—to harness wind power in gale conditions without strain. For avian creatures born into a giant, hemispheric wind tunnel, a howling storm is a lullaby.

The Stormy Petrel and his long-suffering officers might have been poorly prepared for epic Antarctic gales, but they brought with them a fashionable enthusiasm for tracking storms at sea. Meteorology had been a latecomer to respectable science. Well into the nineteenth century, weather record-keeping was considered a hobby for clergymen and cranks. But extreme weather events—storms and hurricanes, in particular—were costly to Atlantic shipping, so momentum built for systematic investigation.

In the summer of 1831, almost from thin air, a paper appeared in the *American Journal of Science and Arts* that helped launch meteorology as a legitimate science in the United States. In the words of meteorology historian Eric Miller, the flurry of weather research that followed the landmark 1831 paper brought "more progress in a decade than had occurred in the preceding millennium."

In his "Remarks on the Prevailing Storms of the Atlantic Coast," a New York mechanic and unknown amateur named William Redfield dismissed all prevailing theories of storms: that they were caused by changes in electricity and temperature, or were mere congregations of clouds pushed by the prevailing winds. Through analysis of recent destructive Atlantic nor'easters, Redfield proposed that these violent events were "progressive whirlwinds," or rotating low pressure systems. He recalled traveling with his son through New England in the aftermath of a violent storm, and observing trees a few miles apart blown down in different directions. This had produced an epiphany in Redfield's mind: winds within a storm system operated independently of the direction of the storm itself, and blew in a counterclockwise direction. What appeared to be a meteorological chaos—winds from everywhere—was in fact a tightly organized phenomenon capable of analysis.

Redfield's theory made a profound impact on the Atlantic maritime community, and no less on its governments and intellectuals. The progress of disastrous storms heretofore assigned to providence could be tracked and even *predicted*. Pioneer British meteorologist William Reid championed Redfield's theory in his influential book *Law of Storms*, published on the eve of the Ross and Wilkes expeditions' departure for the Antarctic, while a young undergraduate at Oxford, John Ruskin, grasped the planetary possibilities of the moment. It was the mission of the meteorologist, the budding critic wrote, to "trace the path of the tempest round the globe." But he could not do so alone. The challenge of modern meteorology was its synoptic imperative: data must be collected from across the globe, involving a legion of coordinated observers. The new meteorological societies of Europe and North America must operate as "the moving power of a vast machine" of science. A worldwide community of weather watchers and storm trackers would—by the aggregation of their data—belong to "one mighty mind . . . one vast eye" of knowledge. From his Christ Church dorm room in 1839, the twenty-year-old Ruskin anticipated our world of orbiting satellites, climate models with a million data points, and the Weather Channel. A vast machine, indeed.

Ruskin published his thoughts on the new meteorology even as the US Ex. Ex. churned southward toward Antarctica, into latitudes conspicuously bereft of weather data. Every officer belonging to the expedition was steeped in the writings of Redfield and Reid and was eager to test their theories, notably that storm winds in the Southern Hemisphere rotated in a direction opposite to those in the north. They could not have known that their opportunity to play their role in Ruskin's "vast machine" of meteorology would come the following year in a form more violent than any thought possible, along the stormiest stretch of coast on Earth.

At the outset, however, like so much else about the US Ex. Ex., prospects for meteorological discoveries seemed bleak. Wilkes saw himself as a pioneer weatherman, but not in the modern spirit of Ruskin. He hoarded all data for himself. When the scientists of the fleet—including James Dana, later America's preeminent geologist—approached their commander with a proposition to collaborate across the fleet on a much-needed weather map for the South Seas, the Stormy Petrel sent them packing. Southern Ocean synoptic data would have to wait for the Heroic Age explorers, and the IGY. In the world according to the US commander, the "one vast eye" of meteorology was cyclopic—his own.

‡ 6 ‡
Madame D'Urville's Letter

Recently escaped from the Weddell Sea ice pack, D'Urville's *Astrolabe* and *Zélée* skirted north along the western tip of Chile in March 1838, chasing the remains of the southern summer. They sailed alongside vast Cordilleran glaciers. The sky was pure, the ice dazzling, so that only the relative warmth of the air signaled to them that they were not still in Antarctic waters. After a spell in Talcahuano, they sailed north again to Valparaiso, where they were met by a mail boat. The officers and men crowded the deck, while D'Urville's heart drummed inside his chest. He distributed a hundred letters without seeing his name. Then at last, at the very bottom of the sack, he made out a scrawled address in his wife's hand. He guessed its contents even as he opened the paper with trembling hands. Madame D'Urville's letter is kept at the National Library in Paris. The paper is, literally, stained with tears:

> My friend, why are you not here beside me, alone, isolated, without support, without help against my despair, overwhelmed by my sorrows, when all my memories assault me hearing the piercing cries of my children. . . . Oh, my Emile. In the middle of the street the cholera struck. His face contorted, he was seized by diarrhoea, by vomiting, everything went to the head. He gave out heart-rending cries. His eyes stared vacantly, he tore and bruised his head. He was so

> beautiful. They took him from me. His hands so beautiful. When you receive this letter you will have finished your work in the ice. You could return now, yes? It is my sole desire. Glory, honour, riches, I curse you, you have cost me too much.

D'Urville could not participate in the usual formalities on entering the port of Valparaiso. Retreating to his cabin, he wrote a long letter to Adélie. He poured his grief for their dead son Emile onto the page. He cursed the day he had ever undertaken this Antarctic voyage. But he could not agree to her plea that he return. His mission must continue. She would keep close watch over their surviving child Jules, a prize-winning student at the *école*, and together be strong without him.

The fateful letter from Adélie on his desk, D'Urville felt the tides of more general misfortune close in on him. Even in this busy port—a hive of seafaring gossip—he could uncover no information regarding the American polar expedition. In the Antarctic, not a day had passed but he thought he might encounter one of Wilkes's fleet patrolling the ice pack or, worse, catch sight of the stars and stripes at the rim of the southern horizon. But nothing. Now he wondered whether the intelligence given him in Rio was wrong. Had the Americans not come south at all this summer? It was baffling.

D'Urville was preoccupied, too, with his English rival James Ross. His first social invitation in Valparaiso had come not from the shore, but from a British frigate moored alongside them in the bay. The captain appeared to know exactly who they were and was eager for news of their south polar campaign. D'Urville could sense the English officers' wariness during formal introductions. But the moment they learned the *Astrolabe* and *Zélée* had reached no farther than 64° south, far short of Weddell, the mood in the captain's cabin relaxed dramatically.

D'Urville watched in silence while his officers gave full vent to their disappointment. The ice pack, they explained to the English, was so much farther to the north than they had anticipated. It trapped them. The men had suffered terribly. But they had charted the ice and several hundred miles of unknown peninsula coastline south of the Shetland Islands. If only conditions had favored them, their commander would have led them much farther south, perhaps to the pole itself. D'Urville cringed as the Royal Navy officers fell over themselves to sympathize. Their French friends must not chide themselves, they said. It had been a noble voyage, and these charts were a credit to French navigation. They must not think their Antarctic campaign a complete failure for having not beaten Weddell. Everyone knew the whaler had been lucky.

The smiling condescension of their British rivals had been painful enough. But this was, it turned out, only the beginning of their humiliations. D'Urville had left valuable men behind him at Talcahuano: two dead, another six too ill to move, two malcontents discharged, and eight deserters. But when he explained his dilemma to Captain Duhaut-Cilly of the French ship *Ariane*, who was duty-bound to oblige him, Duhaut-Cilly looked embarrassed and told him none of the crew of the *Ariane* could be spared. The reasons for the captain's unheard-of defiance soon became clear. The *Astrolabe* and *Zélée*—and their commander—were in disgrace in Valparaiso. A malicious report had traveled from Talcahuano that the French Antarctic mission had achieved less than nothing. That Commander D'Urville had not even dared to enter the Straits of Magellan and had fled like a coward at his first sight of ice.

D'Urville guessed the source of the slander lay with his own compatriots: the French captains who greeted him at Talcahuano. Not for the first time, he felt himself the victim of the French naval character, which routinely placed professional jealousy above patriotism. Had a British mission blundered in the way of these false reports, every Royal Navy officer in the Southern Ocean would have seen it as his duty to suppress the facts in order to preserve their collective honor. Meanwhile, the French tore each other to shreds.

D'Urville's officers, less habituated to treachery, were indignant. They spread like an army through Valparaiso, charts and logs in hand, buttonholing every officer they met to lay out the full history of their two-month tour in the ice—the horror, the heroism, and the impenetrable pack. Meanwhile, D'Urville sent a majestically worded official report, extolling their efforts, to the minister in Paris and had the text circulated in Valparaiso. Before the week was out, Duhaut-Cilly was back in D'Urville's cabin, overflowing with apologies, and offering all the men he needed to fill his list. Several of *Ariane*'s crewmen, their taste whetted for polar glory, had already volunteered.

All D'Urville's doubts and calculations now began to coalesce around a new plan. He would not return to the Antarctic next summer—January 1839—despite the king's wish for the French flag to be planted at the South Pole as soon as possible. After their recent ordeal, followed hard by Adélie's letter, he knew he personally did not have the strength of mind or body for a new campaign on the ice. Neither did the officers and men of the *Astrolabe* and *Zélée*. This had been made clear to him in a thousand ways.

Instead, he would take them west into the Pacific, to the islands he knew well from his two previous circumnavigations—his tropical comfort zone. There they would chart coastlines, study the local languages and customs, and gather shells, plants, and animals for the National Museum's collections. There would be fresh fruit, sun, and rest for his demoralized crews. He would tell no one that their Antarctic plans were merely postponed, not canceled.

The unhappy disadvantage of this plan was to disoblige the king and risk permanent disgrace. Their year's sabbatical would leave the door wide open to the British and Americans, who would surely make their bid for the pole in early 1839. Since their arrival in Valparaiso, D'Urville had barely slept for grief over his son, while worrying about the Americans occupied much of his daylight hours. If the Wilkes expedition was stymied in the ice, as his had been, he would rally his officers for their second attempt on Antarctica, this time setting out from Tasmania, on the far side of the Pacific. Few ships other than whalers had ever sailed southward from the port of Hobart. It would be the crowning triumph of his career if the true path to the South Pole lay at that longitude.

But D'Urville was too pragmatic to be reassured by optimistic projections, especially his own. In choosing to bypass Antarctica for a yearlong cruise in the Pacific, the brutal truth was this: if the Americans raised their flag at the pole that season while he was drinking coconut juice on a beach in Tahiti, nothing could redeem him. In that case, he vowed he would not return to France at all, even for the sake of poor Adélie. Rather than face that shame, he and his men would circumnavigate the globe indefinitely, around and around, for a watery eternity.

Giant penguins, we are told, once migrated west with the current across the ancient Pacific from their strongholds in the Antarctic Peninsula to the nascent beaches of Tasmania and New Zealand. Millions of years later, through the southern winter of 1839, the *Astrolabe* and *Zélée* made the same trek from the American Antarctic Peninsula to Australia, cruising the Pacific islands to the South-East Asian Archipelago. Toward the end of that journey, a single barrel of contaminated water loaded from a Sumatran dock almost brought Dumont D'Urville's Antarctic ambitions to an inglorious end.

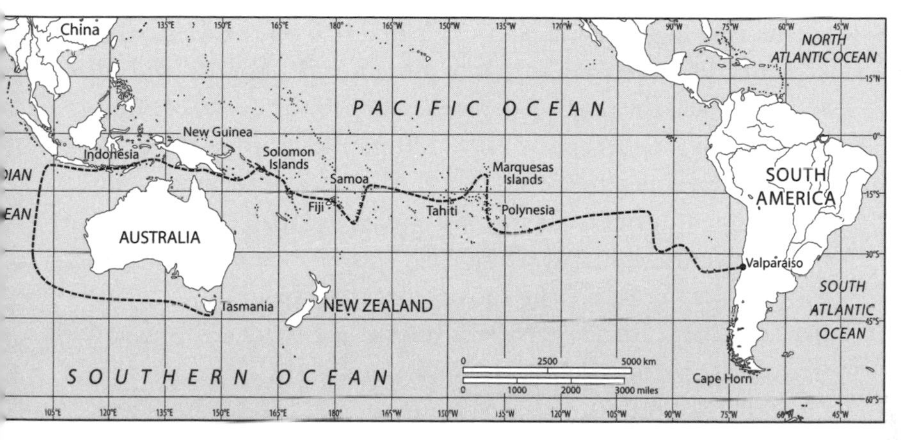

Fig. 6.1. Instead of risking a second polar campaign in the summer of 1838–39, D'Urville led his expedition on a broad-ranging tour of the South Pacific, which he had explored twice in the 1820s.

For two years, the crews of the *Astrolabe* and *Zélée* had endured the extremes of climate across the Southern Hemisphere: the ice and towering waves of the polar circle, burning heat in the Pacific, and the humid island coasts of Southeast Asia. They had outlasted frostbite, scurvy, fevers, diarrhea, blistering sunburns, and a wretched, unvarying diet. Through all this, the basic good health of the remaining crews was as it had been the day they left Toulon. In fact, their cheerfulness had increased in recent months as they checked their mental calendars: day by day, the sight of home and reunion with their families loomed ever nearer. Then, without warning, the horror of dysentery struck.

It is a cruel situation for a sailing ship a thousand miles from any port, on a becalmed sea, to have men sick and dying in numbers below deck. It is twice as cruel, by simple arithmetic, for two identical ships, on identical courses, to be at a standstill on an uncaring ocean while an epidemic ravages her crews. On the silent decks of the *Astrolabe* and *Zélée*, adrift off the west coast of Australia in November 1839, a nightly ritual unfolded in eerie parallel. The starboard watch wrapped the most recent of the dead in his sheets, then lifted him quietly along the deck to a porthole open to the sea. There, they placed the body on a wooden board, and tied a melon-sized cannonball to each bare foot. Once the tortured white flesh was washed and the uniform jacket buttoned, the canvas sheet was sewn up over the matted head. Officers gathered, but no last rites were spoken (for the sake of morale). Sometimes, weirdly, the sea would rise up at that very moment as if to claim the dead, as the body slid headfirst into the starlit swell of the Indian Ocean.

Burials were conducted at night, to spare the crew of the sister ship the sight of the latest victim's chute or, worse, of a shark thrashing the body to bloody pieces. By day, a white sheet stretched out on deck between two oars signaled another loss, while for the next nighttime burial, the ship in mourning would tack discreetly away from her companion, as both sign and screen for the fatal truth. One less man for the Antarctic.

The stricken French corvettes limped into Hobart in the last week of November 1839. Sir John Franklin, the governor of Tasmania (and, later, captain of the ill-fated *Erebus* mission to the Arctic) brought a package of correspondence for the exhausted commander. A brief look at the letter from his wife told D'Urville that her grief, so savage in her first note at Valparaiso, had mollified somewhat. More worrying were the letters from his navy friends who informed him bluntly that his first attempt on Antarctica two summers ago had made no impression on the French public. Nor had their travels since. The expedition was in danger of being entirely forgotten. Bitter as D'Urville was to hear it, the information did him one important service. If he had questioned for a moment his resolution to set out again for the South Pole, those doubts were now erased. For a celebrity explorer like D'Urville, public indifference was a fate worse than death.

News of the imminent second campaign to the Antarctic was met by the French expedition doctors with dismay and disbelief. In the hastily renovated hospital beside the river in Hobart, they pleaded for more time for the invalids. But there was *no more time*. The Antarctic summer was ruthlessly brief. January and February offered the only possibility for above-freezing temperatures and a passage through the glacial belt.

To succeed in the Antarctic, D'Urville needed fresh men to make up for their recent shocking losses. On December 28, three days before D'Urville's deadline to sail, his second-in-command, Captain Jacquinot, produced his Hobart "recruits." D'Urville was appalled. Of the twenty men milling on the deck of the *Zélée*, only a handful appeared seaworthy. The most promising were several deserters from French whalers. A few English sailors, of doubtful background, would also do. But the rest were criminals or idiots, who looked lost holding a rope. One was violently insane. Few had provided their real names. How many English mothers could there be to name their sons "William Watson"? These were fugitives willing to risk their lives in Antarctica rather than face retribution in Tasmania. Not even D'Urville's desperation could match theirs.

The months-long, gut-wrenching question of whether D'Urville would have sufficient numbers to man two ships to the South Pole came down to the final hours. The doctors complained bitterly when D'Urville ordered a full row of feeble patients from the ward to be transported back on board. He promoted several inexperienced midshipmen to acting lieutenant. These, when added to the French deserters and several English escapees, brought him to the magical number sixty-five, the minimum quota of hands—officers and men—to work each ship. He had done what was necessary. History must judge.

And so it was, on the morning of New Year's Day 1840, after a stiff farewell from Governor and Lady Jane Franklin—close friends and compatriots of D'Urville's British rival—the two French corvettes raised anchor and made their way slowly along the River Derwent back toward the Southern Ocean from which they had come. A contrary breeze allowed the departing sailors one last opportunity to admire the steep wooded hills along the Derwent and to inhale the crisp eucalypt air of New Holland. To the south, an odorless, unvegetated, uncertain destination awaited. The waving handkerchiefs of the sailors' two-week sweethearts lingered in view. No one was in a hurry to go below.

Mere hours after their departure, Ernest Goupil, the popular expedition draftsman—and last of its dysentery victims—expired. The following day, in pouring rain, he became the first and last French artist to be buried with military honors on Australian soil. His body was brought, on a horse-drawn carriage decked with the French flag, to the cemetery, where a stone memorial had been hastily erected to the dead heroes of the French South Pole Expedition of 1838–40. Goupil's name, newly etched, brought the number on this roll of honor to thirty-three. The local stonemason, who knew something about ships and their southerly expeditions, had discreetly included several blank panels on the little cenotaph for when room for further names should be required.

⸸ 7 ⸸

Ross Falls Behind

While Dumont D'Urville beat the odds to launch a second assault on the South Pole in January 1840, with the Americans closely astern, the British expedition had only just arrived in the Southern Hemisphere. Cape Town was their launch point into the Southern Ocean maelstrom. Below deck on the *Erebus*, Joseph Hooker bunked amidships with a group of junior officers. On days of calm, the crockery for breakfast, afternoon tea, and dinner rotated smoothly on the sideboard, alongside silver-plated serving dishes and a crystal decanter. But seas in the Southern Ocean were rarely calm, and these luxury items—for which the officers had laid out personal funds—were now mostly stowed away.

Because their berths were situated crosswise, Hooker and his messmates suffered the most from the constant rolling of the Southern Ocean, where a wave might circumnavigate the globe without ever meeting a beach. Meals were ruined by the inevitable aftermath: hours lying in a bunk with the sensation, twice a minute, of being pitched downward while the contents of one's stomach sloshed upward to the thorax. Well before they crossed the equator, they had exhausted their supply of fresh ham, potatoes, and greens and were thrown upon ship's provisions—rations of salt beef, salt pork, and pea soup in a never-ending, repulsive circuit. Wine consumption increased proportionally until it, too, had to be rationed.

With his last ten pounds from his father, Hooker had furnished his cabin with curtains, a rug, and a work desk covered with an oilcloth for dissections. Above his desk, he hung sketches of his family he had made on a walking tour in Invereck. In one, he had drawn his beloved brother William with his wife and their dog before a backdrop of the Scottish Highlands. In another, he had captured his little sister Mary sketching with William's cat next to her. The whole family climbed mountains and rendered landscapes with Victorian vigor. Once in the Trossachs, Joseph and William had walked a hundred and ten miles in three days with only a hunk of bread in their pockets. During *Erebus*'s four-year voyage, when Joseph's thoughts turned to home as he lay on his cot, he preferred to imagine himself with his family in the wilds of Invereck more than in the poky drawing room on Woodside Crescent in Glasgow where he had in fact grown up.

As it turned out, Hooker spent little time at the desk in his cabin. Once Captain Ross realized that Sir William Hooker's young son—whom he had taken aboard as a favor to that distinguished person—was not a worthless gentleman tourist but actually dedicated to his profession, he offered him a desk in the captain's cabin under the stern window with its streaming light. He was permitted to keep his microscopes and specimen papers in the desk drawer, and an entire cabinet was made available for his plant specimens. Besides this, the floor of the captain's cabin was strewn with boxes, buckets, and bottles by the dozen for the growing collection of marine samples. Hooker had trained as a botanist but here on the open ocean he was largely deprived of plant life and became, perforce, a marine biologist. It was not long before he realized that the captain—who had ambitions in that line—had recruited him as a marine research lackey.

From the desk at the broad stern window, Hooker could keep an eye on the two towing nets lowered in the ship's wake. Their briny contents were brought to him several times a day. By the time they reached the African cape, he had sketches of more than two hundred species of mollusks and crustaceans, drawn with the aid of the captain's superior microscope. Because the best catch was to be had in the evening about nine o'clock, he and Ross often stayed up working by candlelight until three in the morning. The captain absorbed himself in the complicated working of his daily magnetic observations, occasionally coming over to see what new delicate wonder of marine life Hooker was hurriedly rendering before rot set in.

Early in the voyage, in the tropical waters off Brazil, they came across large, shining patches of pyrosomes, what the sailors called sea pickles. Peering over the side at night, he could read a book by their silvery light. One evening in Simon's Bay off Cape Town, the net disgorged no fewer than thirty different creatures—shrimps, crabs, worms, and sponges, as well as picturesque corals. Hooker got no sleep at all that night. On the long, stormy haul from Kerguelen Island to Hobart, they sailed through meandering lines of mollusks, miles long, twisting across the water's surface like giant brown sea snakes. The ubiquitous seaweed was the best storehouse of all. From it they extracted innumerable tiny, squirming forms they couldn't name, to preserve in jars until such time as they could be named.

In accounts of his shipboard life to his botanical father, Joseph was careful to minimize his marine studies, but he found himself fascinated by the miniature oceanic world disclosed to him beneath his microscope: little floating shells with tiny appendages for propulsion, some like wings, others like feet, sails, bladders, and fins. At night, the wake of the *Erebus* lit up with the electric glow of a million animalcules. The flashing lights appeared to attract tiny shrimps—the ubiquitous krill. It was no effort to collect thousands of these in one pull of the net. At Hooker's insistence, they began a catalogue of seaweeds, so that the steward grew tired of finding his captain and the learned young gentleman shoulder deep in green, leathery weeds, put in salt water buckets to be drained, bottled in spirits, or laid out in great messy strands on the table to dry.

This working relationship was a source of satisfaction to both men, though more so to the captain. Ross thought Hooker's botanical collections from the Atlantic islands the best thing he had ever seen, to which Hooker's unspoken reply was that the captain clearly had never known a real collection. Regarding scientific booty—which certain philistines at the Admiralty had argued should play no part in the expedition whatever—naval rules dictated that all specimens, and observations on them, belonged by right to the Admiralty. The question of "personal" collections was left unresolved, since it would be tactless to raise it. At Chatham docks the previous summer, Hooker had shown his slight regard for naval protocol by striding into Ross's cabin to request a berth to Antarctica as the expedition's naturalist, with no real qualifications to boast of. He had shown the same nonchalance for Admiralty rules at Saint Helena, where he secretly shipped a crate load of botanical specimens to his father at Kew, along with a fiercely annotated duplicate of his notes. Now, when *he* looked at the growing crowd of marine specimen jars in the captain's cabin, he did so less to admire them as to wonder whom they belonged to, and who would win the credit of publishing their contents.

During the eastward leg from Kerguelen Island to Tasmania—so many months out of England—Joseph Hooker had anticipated, in dreamy detail, the moment of their arrival in Hobart when he would open his first letters from home. When the time came, however, the mail brought only a piercing shock. The letter from his father was bordered in black and addressed "To My Only Son." William, his older brother, had died after a brief illness. In his grief, the entire wonder-filled narrative of Joseph's South Sea adventure turned to ashes, since it was to William that he had always imagined himself telling it. For the length of their stay in Hobart, he refused all invitations to Government House where—with the encouragement of Lady Jane Franklin—the expedition's officers entertained the daughters of the best families of Hobart. The young naturalist took refuge from society in the mountains behind the town, where he exhausted himself in the assembly of a floral collection so original and complete that it has served as the basis of all botanical science in Tasmania since.

But his brilliant exertions were not apparent to his father, who had sorted through Joseph's first shipment from Saint Helena with rising frustration. The specimens were so poorly preserved after their long journey that the meticulous accompanying notes were mostly useless. It was a further crushing blow for Joseph Hooker to read a letter from his father admonishing his lack of effort. After these worst weeks of his young life, Hooker found himself wishing for the ice, for the therapeutic effect of pure, numbing cold.

James Ross's bad news had come not from home, but from his old Arctic shipmate John Franklin. The *Astrolabe* and *Zélée*, the governor told him, had arrived in Hobart the previous spring in an appalling state. Dysentery had struck them in the East Indies. D'Urville had lost dozens of men, including some of his best officers, and for a time it seemed doubtful whether the Frenchman had sufficient numbers to crew even a single vessel for a second bid to the South Pole. But D'Urville's determination and professional skill won through, and the French corvettes had duly set sail for Antarctic waters.

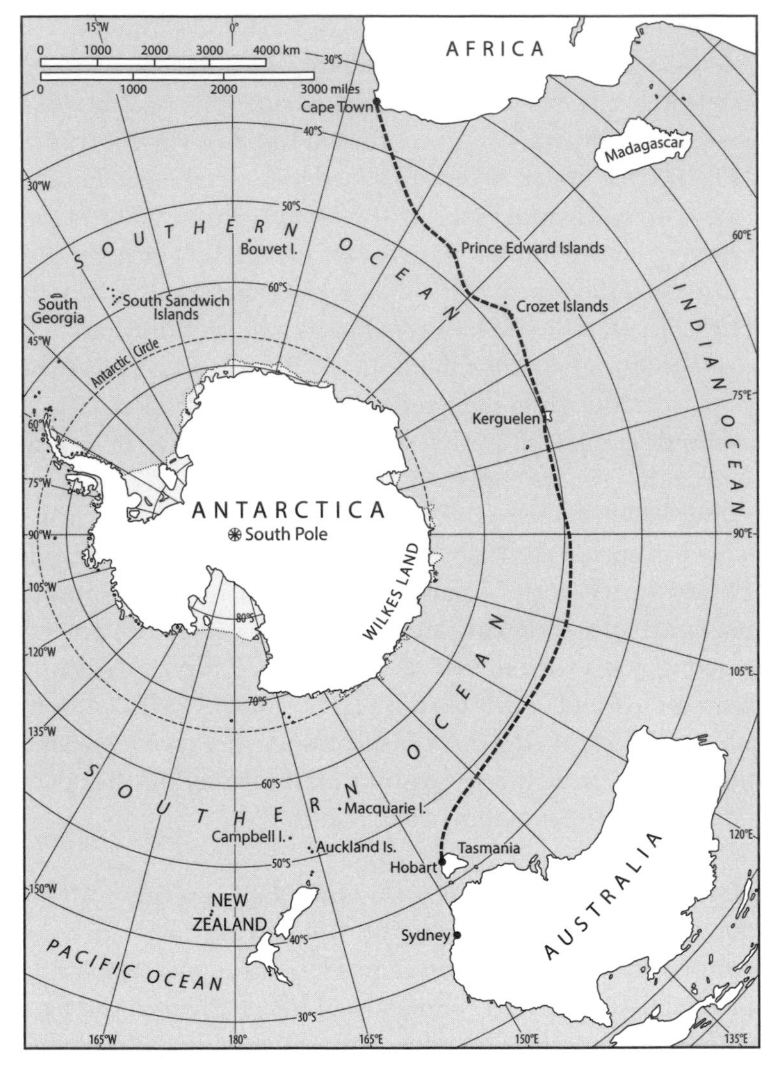

Fig. 7.1. James Ross chose an alternative route to Antarctica from his rivals', crossing the Indian Ocean via Kerguelen Island to Hobart, Tasmania, where he arrived in August 1840.

Worse still, from Ross's point of view, the French commander had confided to Franklin his intention to follow exactly the southerly course Ross himself had planned for the *Erebus* and *Terror*. Ross's rage at his French rival for preempting him ran deep. During their Hobart sojourn, not even Lady Jane Franklin's attentions to him as "the handsomest man in the Royal Navy" and the continual round of social hilarity could relieve his gloom. In fact, these only added to his worries. Captain Crozier of the *Terror* had promptly fallen in love with Lady Jane's companion Sophie Cracroft, a young woman who spoke French and dazzled his officers without mercy. She had refused poor Crozier—who was no ladies' man—while intimating her preference for Ross himself. The episode seemed to have thrown his old friend into a deep depression, at exactly the time his steady cheerfulness would be most needed.

James Ross was a soldier of the queen. To his imperialist mind, the British claim over the polar antipodes could never be in question. His own years of service in the Arctic wastes, the British presence in Tasmania and New Zealand, and the feats of James Weddell all certified these southern waters as British—his to explore by right. But D'Urville had ignored all this, and a British commander could not follow in a Frenchman's wake. So Ross tore up his plans. From the outset, he had been a distant last in the tri-nation polar campaign, and only drastic measures offered a chance at victory. He would sail southeast instead, into the complete unknown, where no Frenchman had ever gone. Nor, once meeting with pack, could he afford to turn back. This circumstance, as it turned out—Ross's desperate last throw of the dice in Hobart—would determine the outcome of the historic first race to the South Pole.

PART THREE

⚓

Triumph

‡ 8 ‡

Seas of Grass

A week's sail out of Hobart, passing 60° south, the *Erebus* and *Terror* met with their first icebergs. They noticed a northward current drawing them away, like a warning tug. At the so-called Antarctic Convergence, where the temperate waters of the north become involved in the polar current—the most powerful in the world—the temperature differential of the sea, at surface and depth, began to fluctuate wildly. They became aware of a great submarine turbulence visible in the whorls and eddies of the surface water, and an orange-brown carpet of algae enveloping the *Erebus*.

Joseph Hooker had expected the daily catch in their tow nets to diminish as they headed south into the cold unknown. Instead, they came upon a marine playground the likes of which he had never seen. Whales of all kinds paraded about the ship—the fin, the humpback, and the stupendous blue whale—grazing on the algal film of the ocean's surface like cattle on a meadow. The birds—broad-winged albatrosses and zooming petrels—congregated in huge numbers, some skimming and scooping up their catch, others settling on the waves and dining in. On deck, the tow nets unloaded the highlights of their meal: sea butterflies and snails, tiny octopuses, paper-thin nautiluses, crustaceans, and diatoms, barely visible even under the captain's superior microscope.

Diatoms, a ubiquitous plankton, serve as the bottom of the oceanic food chain. Scattered on the ocean, in filament chains,

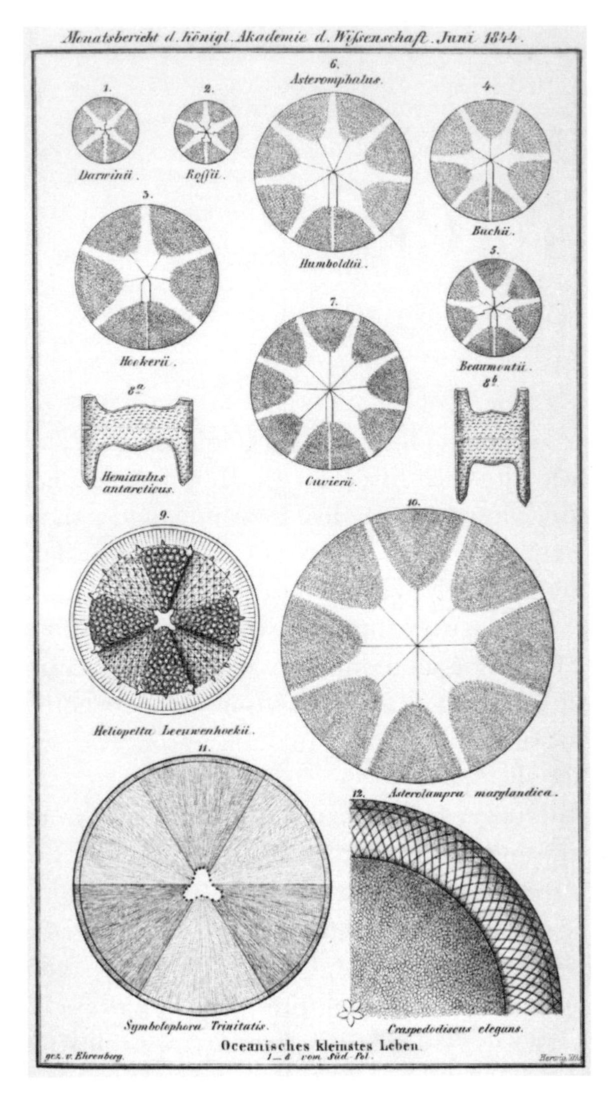

Fig. 8.1. Illustrations of Antarctic diatoms, based on sketches and samples collected by Joseph Hooker. With identification of seven new genera and seventy-one new species, Hooker's research on diatoms, microscopic "jewels of the sea," ranks among the most significant scientific contributions of the Victorian polar voyages.—C. G. Ehrenberg, *Verhandlungen der Königliche Preussiche Akademie der Wissenschaften, Berlin* (1844). The Royal Society.

the diatomic mass assumes a distinctive orange color, like burnt sienna. If the ship's course was leisurely enough—less than two or three knots—Hooker could reach a fine muslin net over the side and draw up a bounty of diatomic plankton. Filtered through fine-weave paper and placed under the microscope, the granular structure of the vegetable cell, now a crystalline yellow, was clearly visible. And it was alive.

The diatoms marked the rim of the pack ice and the underside of icebergs with a brown and reddish stain. Where the ocean surface congealed with cold, in the first stages of freezing, the trapped diatoms, in their millions, turned the ice an ocherous color. Hooker discovered brightly colored diatoms in the stomachs of shrimp-like salpa and krill. He found diatoms in penguin guano. When they dredged the ocean flood in the vicinity of the ice, an idea of the vast sedimentary deposits of diatoms, taking place over untold eons, was impressed upon him. Placed in a water glass, the samples of the freezing mud, typically white or green, turned cloudy like milk. Returning hours later, Hooker would find the diatom cells sunk to the bottom of the glass, enclosed in their indestructible silica walls.

It was "a wonderful fact," Hooker wrote in his journal, how this great invisible harvest of diatoms formed the base of all marine life in the Antarctic. The *Erebus* and *Terror* arrived at the Antarctic Convergence in the days after Christmas 1840, approaching midsummer, in the middle of the annual feeding frenzy of the Southern Ocean, when the brief pulse of solar energy in the Antarctic summer reanimates the dormant food web. As the pack ice retreats southward, bacteria-rich meltwater nourishes diatoms and other algae, creating blooms feasted on by krill, which in turn feed the fish, birds, and whales. Beneath the frenzy, however, Hooker observed a distinct orderliness.

The diatom-krill-whale triad is among the shortest food chains in the world, remarkable since its apex predator, prior to South Sea whaling, constituted the largest mammalian stock ever to have existed. More wonderful still was to imagine the sheer volume of krill and diatom stocks required to sustain them. At the convergence, no single whale group monopolized the feast. Hooker witnessed the departure of the blue whale pods and the arrival of the fin whales. Then the humpbacks took their turn at the trough, resurfacing vertically with their plated baleen mouths open to engulf the krill.

Within each pod, too, rigid protocols applied, to maximize survival. The pregnant females arrived first at the feeding ground and left last, followed by the young males, females with calves, and mature males. Each whale consumed up to eight thousand pounds of krill a day during their two-month long stay at the polar front. Every year their numbers were diminishing on account of the human thirst for whale oil, but even in 1840, with hundreds of American whalers in the South Seas, the great baleen whales collectively consumed more than a hundred million tons of krill during the polar summer.

Hooker was under specific instructions—from the European science czar Alexander von Humboldt himself—to take special care with the diatom collection in Antarctic waters. He was to deposit samples in tiny phials provided to the expedition and forward them to Professor Ehrenberg in Berlin for identification. In the manuscript copy of *Voyage of the Beagle* that Hooker kept under his pillow, Darwin told the story of a windy afternoon off the coast of Africa, when a squall deposited a thin, brown layer of organic life across the *Beagle*'s decks, airlifted from the desert. Ehrenberg had identified these as diatoms, the microscopic food source of the world, even venturing to identify their sexual organs.

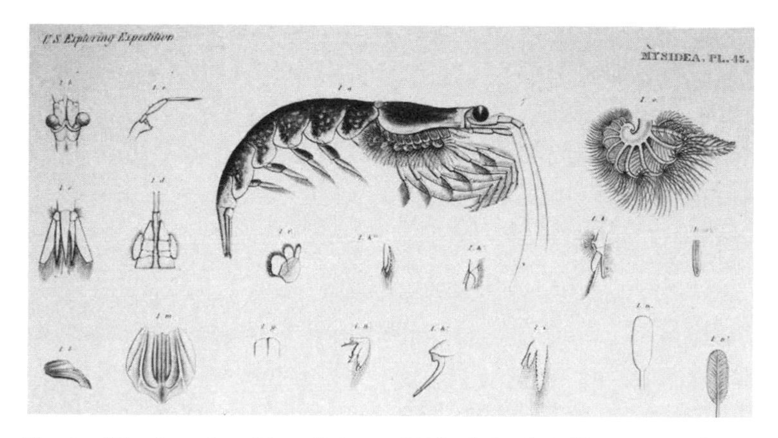

Fig. 8.2. The American Expedition published the first illustration of krill, foundation species of the Antarctic food chain.—James D. Dana, *United States Exploring Expedition,* vol. 14, *Crustacea* (Philadelphia: C. Sherman, 1855). Biodiversity Heritage Library/Smithsonian Libraries.

But working late by candlelight aboard the *Erebus*, Joseph Hooker first intuited what has been called the greatest discovery of the nineteenth century in biological oceanography. Diatoms were *plants*, not animals—a microscopic vegetable. To Hooker the aspiring botanist, the Southern Ocean had seemed like a desert, occupied only with animal life. The ocean teemed with shrimp and mollusks and crustaceans, preyed upon by seals and penguins and birds in great variety: all carnivores. But here was the vegetable base of Antarctic life, the herbage of the ocean on which all the charismatic animalia of the higher orders depended, including the great whales. Hooker's insights did not end there. Their rich, silicon character told him that diatoms were also involved with the atmosphere above, exchanging gases, and purifying the air. Diatoms were the grass of the sea, and its leaves and trees—a vital, microscopic agent in the great business of polar nature.

Interlude: The Antarctic Convergence

Jim Kennett, a central figure in the early history of paleoceanography, was the same age as Joseph Hooker when he first saw Antarctica. And, like Hooker, his destiny as a scientist led him to the Antarctic Convergence and the study of diatoms—to uncover extraordinary secrets of Earth's and Antarctica's glacial history. A precocious geology senior from Victoria University in Wellington, New Zealand, Kennett was offered an unusual study abroad opportunity in the spring of 1962: an expedition to the Darwin Glacier in the Transantarctic Mountains, which had never been mapped, where no human beings had, in fact, ever been.

The geologists flew into Scott Base, on McMurdo Sound, in December, the beginning of the summer research season, then south to an area named Brown Hills, adjacent to the Darwin Mountains. To the northwest lay the famous Dry Valleys, a Mars-like terrain containing the oldest exposed rock on Earth. In the Dry Valleys, it had barely snowed in a million years. Brown Hills attracted the team from Victoria University for that reason: bare rocks are a precious commodity in Antarctica.

They arrived with ambitions to map the area by traversing the Darwin Glacier on foot, man-sledding their stores in the heroic style of Scott and Shackleton a half century before. But once on the ice, they instantly gave up this notion. Even this single Antarctic glacier was ridiculously vast, far beyond the capabilities of a party of undergraduates and their professors. So they called in air support. From ten thousand feet, Jim Kennett gazed down at the ice sheet, a great white plateau to which there was no horizon. Though he did not realize it in the moment, his career path as a scientist stretched out before him. There could be no greater physical phenomenon on Earth than this continental ice fortress. It must have shaped the world. Why was it so little studied? Thereafter, Kennett viewed everything he saw through the glaring lens of Antarctic ice.

The richest universities and research institutions in the world are located in Europe and North America. From their northern hemispheric locales, the paleoclimatologists of the mid-twentieth century focused their attention on the Arctic ice ages, when northern ice sheets had spread as far south as England, Germany, and the American Midwest before retreating to their current redoubts in Greenland and the Canadian tundra. The ice age cycle had begun two to three million years ago when, it was assumed, Antarctica had also iced over. Bi-polar cooling had been coincident because there was no reason to believe otherwise.

Enter Jim Kennett, a son of the Southern Hemisphere, to overthrow this complacent assumption. For his doctoral thesis, written at the age of twenty-four, Kennett studied fossilized mollusks across the North Island of New Zealand. These mollusks allowed him to travel even further back in time than the Pleistocene ice ages, back to the Late Miocene epoch, fifteen million years ago and older. The character of these mollusks, and their deposition, indicated that New Zealand had witnessed major episodes of sea-level fluctuation.

One mechanism for sea-level rise was, of course, subsidence of the land, which had been Joseph Hooker's correct theory about the sunken continent of Kerguelen. But changing sea levels in the Northern Hemisphere, since the Pleistocene, were now attributed to the advance and retreat of Arctic ice. Why would the same not hold true in the Southern Hemisphere? Implicit in Kennett's reasoning was the radical proposal that Antarctic glaciation had been initiated millions of years before the ice ages of the north. His New Zealand mollusks had evolved to favor more cold-tolerant varieties. This fantastical ice sheet had not only changed sea levels; it had revolutionized climate and climate-dependent creatures.

Fortune favored brave speculation. Kennett had just embarked on his academic career in the United States when the Deep Sea Drilling Program (forerunner of the ODP) was launched, with the goal of certifying plate tectonic theory through study of the ocean floor. For Kennett—who required no convincing on continental drift—the program offered more enticing opportunities: to study the sedimentary history of the South Seas and to test his theory of a Miocene Antarctic glaciation. The groundbreaking Leg 28 of the DSDP, in December 1973, was to explore the feasibility of deep-sea drilling in Antarctica, in the most atrocious conditions on Earth. Jim Kennett was then booked aboard Leg 29, which would cruise Antarctic waters south of Tasmania in late summer, hoping to unearth details of the ice continent's original creation after its first rupture from Australia fifty-five million years ago.

While he waited impatiently for the Leg 28 and 29 voyages, Jim Kennett returned to his native waters, the Southwest Pacific north of New Zealand, aboard Leg 21. As with all the early deep-sea drilling missions, the principal goal was geophysical, not oceanographic: that is, to learn the tectonic history of the Coral and Tasman Seas. This geophysical monopoly on drilling could be oppressive. When, on their departure from Fiji in November 1972, the team leader suggested that they drill down only to basalt and not bother to core the ocean sediment overlying it, Kennett's response was to the effect of "Over my dead body." So they extracted cores that told a paleontological story stretching from the present day back to the last days of the dinosaurs.

Shirtless on deck in clear tropical waters south of Fiji, Kennett little expected to uncover the secret history of Antarctica. The first three drill sites offered a rich trove of ancient marine fauna but few surprises. Microscopic plankton—diatoms, foraminifera, radiolarians, in all their stellar variations—dotted the sedimentary clay. Here was the base of the southern marine food chain reaching back fifty million years, in an almost unbroken sequence.

The drillship *Glomar Challenger*—predecessor to the JOIDES *Resolution*—then headed southwest to more temperate waters, a few hundred miles east of Sydney, where an intriguing anomaly presented itself. Coring to three kilometer depths, the ancient plankton disappeared for about fifteen million years, spawning confusion in the paleontology laboratory aboard the *Challenger*. In the half-circle section dated to forty-five million years ago, single-celled diatoms were abundant. Then they were gone. Around thirty million years ago, new diatoms and other plankton reappeared, in their modern form, with similar abundance but lower diversity. The ocean waters had cooled dramatically, then warmed again, leveling off at their current temperature.

At the first anomaly, Jim Kennett assumed some tectonic event had erased the record. But when four more sites, from New Zealand's west coast hundreds of miles north to the Coral Sea south of New Guinea, produced the same faunal gap—more missing plankton—he truly began to wonder. The Pacific Ocean east of Australia had been warm from the Cretaceous into the Eocene. Then, at the beginning of the Oligocene, sea temperatures had abruptly cooled, wiping out an entire marine ecosystem and eroding the seafloor. Millions of years passed, heat was gradually restored, and a new and different sea fauna emerged, persisting to the present day.

This "vast regional unconformity" in the South Pacific, and Jim Kennett's explanation for it, drew the attention of the editors at *Nature*, the most prestigious science journal in the world. In a paper published in September 1975—only eight months after the conclusion of Leg 21 (a heartbeat in academic time)—Kennett argued that the remarkable erosion of all deep-sea sedimentary faunal record during the Eocene-Oligocene Transition (EOT) was due to the sudden northward incursion of freezing bottom water currents from a newborn Antarctica. These results from the Coral Sea added to a growing body of paleoecological evidence, from sites across the globe, that the EOT was a time of dramatic cooling in the planet's history. At some point over the subsequent ten million years, Antarctica's break with Australia, at their Tasmanian hinge, was completed. A great new ocean current—of global power and significance—was established. This south polar current circled the new ice continent in its entirety, sealing the cold within its high-latitudinal bands, and creating the largest seaway on modern Earth: the Southern Ocean. Heat returned to the South Pacific and the rest of the world, and with it a new, modern order of marine life.

For Kennett, the absence of plankton remains deep beneath the Coral Sea revealed, in sedimentary code, the story of Antarctica's birth and, with it, the creation of modern world climate—a grand theory, but unproven by any physical evidence from Antarctica itself. With the publication of the speculative Leg 21 paper in *Nature,* the significance of the upcoming Leg 29 mission loomed larger than ever. This voyage would take Kennett to the Southern Ocean itself, to the submarine boundary where temperate and polar waters meet. The stakes were high for a young geologist, barely turned thirty. Kennett's graduate school thesis of early Antarctic glaciation had now evolved into something much larger: a theory of the modern Earth system. A sedimentary core drawn at the Antarctic Convergence would confirm whether he was right or must recant.

At the Antarctic Convergence, the turbulent mix of temperatures and currents produces one of the richest marine feeding grounds in the world, a veritable cafeteria for whales. One hundred and thirty years after the *Erebus* explorers celebrated their Christmas feast with the whales at the Antarctic Convergence, Jim Kennett traveled there aboard Leg 29. He had his own reasons for fascination with diatoms, Hooker's great legacy to science from the Ross Antarctic Expedition.

Diatoms thrive the world over, across land and sea. They cling, in a gelatinous film, to submerged plants, seaweed, or algae, or float together in long, brotherly chains. Collectively, they represent 40 percent of the primary production of the world's oceans, and fully half of the organic carbon buried in the seafloor. Today more than a hundred species dominate the oceans, in mesmerizing variations, some square, some trapezoid, some elliptical. They thrive most of all at the poles, craving the nutrient runoff from melting sea ice. The Antarctic continent is swathed by a deep-sea sedimentary belt of diatomaceous ooze up to two thousand kilometers wide—between 45° and 65° south—a heritage dating back to the Eocene-Oligocene Transition and beyond. Diatom blooms form the grazing grounds for zooplankton and crustaceans, while those that live out their natural term sink slowly after death below the surface waters. Of these, perhaps 5 percent find their final resting place on the seafloor, where they form a rich trough of paleoecological data for the subsea grazing Earth scientist. In the laboratory, the fossilized cell walls of a million diatoms appear as a soft, sedimentary ooze that, when dried, crumbles to powder. Under the microscope, they reveal surprising crystal patterns, hence their nickname "jewels of the sea."

By studying the appearance and disappearance of different diatom species, and their abundance within different layers of sediment, the oceanographer can deduce the sea-surface temperatures of the ancient past, the evolution of deep-sea ocean currents, and the history of ice sheets. The evolutionary turnover of diatom species is relatively short—two to three million years on average—making them ideal biostratigraphic markers. Moreover, the chemical analysis of diatom silica isotopes offers clues to both the history of the global silicon cycle and the diatoms' role in climate-shaping fluctuations in atmospheric carbon dioxide. Residents of

Earth for one hundred twenty-five million years, and uniquely responsive to changes in their environment, the tiny, spindle-shaped diatoms are a vital natural signature for both glaciation and the greater cycles of climate change.

In March 1973, the sedimentary cores drilled by Jim Kennett and the crew of Leg 29 at the present site of the Antarctic Convergence were the first ever to be retrieved in Antarctic waters. Study of the microfossil collection of Leg 29 was historic for another reason: it was the first to apply Nicholas Shackleton's revolutionary new dating technique of isotopic analysis. Through this tool, the Leg 29 diatom deposits revealed an entire new history of our modern Earth, oceans, and climate.

For the first time, the hothouse of the Early Eocene, fifty million years ago, came clearly into view, followed by a gradual temperature decline. But the real climate revolution was still to come. Until about thirty-eight million years ago, a greater Tasmania still clung to Antarctica, connected by a highland range. Then, approaching the Eocene-Oligocene boundary, diatoms reappeared under the microscopes below deck on the *Glomar Challenger*. A marine gateway had been breached, and Antarctica came into existence as an independent continent. Australia, with Tasmania at its southern foot, continued its northward drift, leaving Antarctica sole resident of the pole. The waters of the Indian and Pacific Oceans rushed to unite, creating in the process an entirely new, southern ocean. Deepwater diatom species replaced the shallow. Because the infant Southern Ocean encountered no landmasses in its circumpolar course, it mixed with all the great oceans and, by degrees, began to reshape their character. An increased temperature gradient across the Southern Hemisphere, caused by the glaciating pole, intensified zonal winds, which awoke the oceans from their Eocene sluggishness. The cold bottom waters of the new Antarctic Circumpolar Current generated sufficient energy to drive a new global ocean circulation that carried salty polar water toward tropical seas, which, as part of an overturning conveyor belt, delivered warmer waters to high latitudes. The Gulf Stream, and with it the climate zones of modern Earth, was created.

At thirty-four million years ago—the Eocene-Oligocene Transition—the extinction of warm-zone diatoms and appearance of cold-tolerant successors signified that the Tasmanian gateway had fully opened and glaciation of Antarctica had begun. From their 20°C highs during the Eocene Climatic Optimum, sea surface temperatures plummeted to 5°C, close to modern levels. The new ice sheets, and their fierce winds, delivered bountiful nutrients to the oceans, sparking a phytoplankton explosion. For millions of years, while the passage between the newly separated landmasses was still narrow, the force of the invading current between them was so intense as to erase all microfossil record. Then, in the Miocene epoch, beginning twenty-two million years ago, the diatoms return to view, this time in their fully modern shapes. High polar winds concentrated plankton blooms into belts of superabundant productivity, such as the Antarctic Convergence. In turn, the diatoms—the grass of the sea—drew carbon dioxide from the atmosphere and buried it in the ocean floor, reinforcing the windy cooling cycle. The consolidation of the Antarctic Circumpolar Current certified Earth's transition from hothouse to icehouse, and the symbiotic relation between Antarctic ice, the diatomic production of the ocean, and global climate, became permanent.

But this is not a history of the oceans only. During the Eocene Climatic Optimum, in which the world sweltered fifty million years ago, carbon dioxide in the atmosphere was six times what it is today. As the Eocene progressed, tectonic rearrangement of continents altered ocean circulation, and the planet incrementally cooled. Long before the Arctic and Antarctic were glaciated, ice caps began to form in alpine ranges, lowering the sea level, while grasslands began to take the place of tropical rainforests. Proto-modern grass-eating hoofed mammals replaced leaf eaters.

Exposed coastal rock and grassland prairies combined to energize the global silica cycle. The weathering of rock deposited enormous reserves of silica into the oceans, as did runoff from silica-rich grasslands. In this generous new environment, the diatoms, silicon life forms par excellence, assumed their dominant place in the life cycle and biomass of the oceans, preferring especially zones of high turbulence where nutrients upwelled through the water column and the vicinity of melting ice. Nowhere in the world better satisfied both these requirements than the Antarctic Convergence.

Jim Kennett published his diatomic data from Leg 29 in *Science* (1974) and *Nature* (1976), with Nicholas Shackleton co-authoring the latter paper. In these seminal essays, Kennett was able to provide physical evidence for his theories of Antarctica's birth, the onset of polar glaciation, and the beginning of modern global circulation of the oceans. Some had argued that the experimental extension of the Deep Sea Drilling Program to the Antarctic Ocean was a boondoggle, of interest to only a small outlier group of Earth scientists. Others viewed the program's mission in purely geophysical terms, to extend understanding of Earth's tectonic history. But it's the virtue of good ideas to have unintended benefits. In the work of Jim Kennett and his colleagues, Leg 29 helped launch an entire new field—paleoceanography—and signal a revolution toward a twenty-first century Earth system science that links the study of land, oceans, atmosphere, and biota in an interconnected whole.

For Ewan Fordyce, a scholar of whale evolution and New Zealand compatriot, Kennett's papers had just such a holistic, transformative effect. Modern whales had branched from their ancient forebears at the time of the Eocene-Oligocene boundary, for reasons unknown. The most extravagant development at that time was the emergence of the baleen whales—including the blue whale, minke whale, and sperm whale—which dispensed with teeth in favor of a complex, filter feeding orifice capable of vacuuming

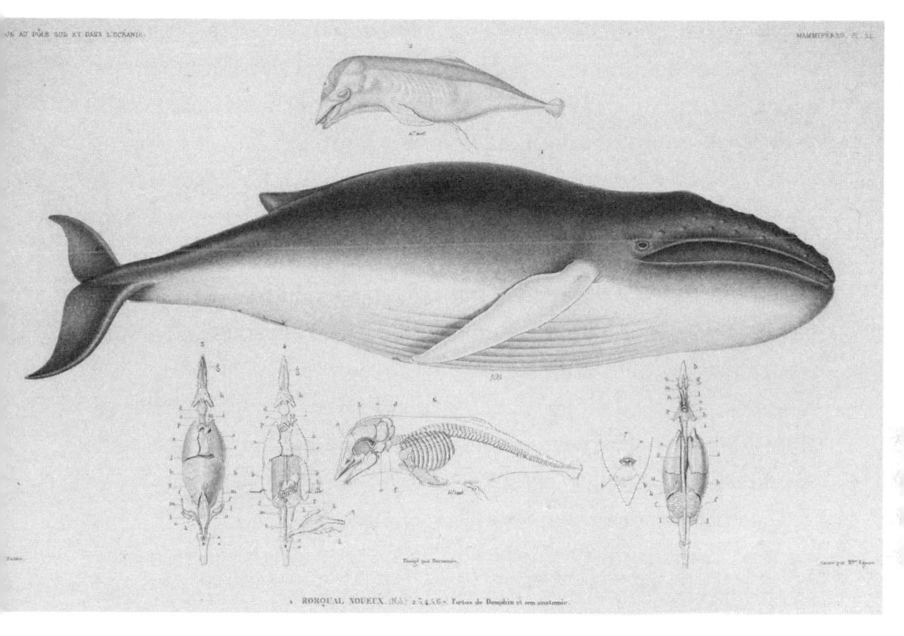

Fig. 8.3. The Victorian polar explorers encountered legions of humpback whales, which feed on krill in polar waters, prior to the period of their near extinction.—Dumont D'Urville, *Voyage au Pole Sud et dans l'Océanie sur les Corvettes L'Astrolabe et La Zélée . . . Zoologie* (1842–53). Biodiversity Heritage Library/Smithsonian Libraries.

up krill by the ton load. The earliest evidence of the baleen whale's existence had been found in fossils unearthed on the south coast of New Zealand.

With Kennett's paper on the opening of the Tasmanian gateway at the EOT, Fordyce was handed a trigger for modern whale evolution. The formation of the Southern Ocean and northward expansion of Antarctic ice had sparked a surge in phytoplankton mass, including the cold-loving diatoms. These diatom blooms offered powerfully selective nourishment to a single crustacean, the *Euphausia superba*: Antarctic krill. Krill, in turn, selected for those whales that could eat them, as well as for size—for those whales that could store sufficient energy to survive the long interval between biotic pulses in the austral summer, and could cruise to the tropics for breeding during the off-season.

A radically shortened food chain proved a turning point for whale dominance of the oceans, ended only by the intervention of humans with their pitiless harpoons. The first humans ever to sail the Antarctic Convergence at this critical longitude—the explorers of the Ross Expedition in December 1840—remarked on the extraordinary proliferation of whales drawn to the rich vegetable blooms on the water's surface. Joseph Hooker—with his gift for systems thinking—grasped the keystone role of the tiny diatoms in the Antarctic ecosystem.

But neither Hooker nor his fellow explorers could know that their captain's secret course had taken them to the very straits where, by its final partition from Tasmania, Antarctica itself was first created a continent thirty-four million years ago, and where, amid the grand global reorganization of land, seas, plants, and animals that ensued, the baleen whale emerged in its mouthy glory. All but the most jaded whalers among them crowded the sides of the *Erebus* to witness this feast of giants, in which was mixed, at some cetacean register, a spiritual joy of return to their original waters.

‡ 9 ‡

How the Adélie Penguin Got Its Name

A full year before James Ross—in January 1840—Dumont D'Urville's discovery ships had passed the Antarctic Convergence several points to the west. Where Ross's course would lead him by pure chance to the world's southernmost sea, polar volcanoes, and a floating ice plateau beyond all imagining, the French expedition would instead draw beneath the loom of the East Antarctic mountain coast—a place arguably less like planet Earth than anywhere else, but whose vast ice cap holds the key to humanity's future distribution on its surface.

Farther north than D'Urville anticipated, the *Astrolabe* and *Zélée* fell in among giant icebergs as flat as billiard tables. He could not imagine such monsters generated by the sea, so reasoned that land must be in the offing. The air was biting cold, but the sea turned peaceful, like a lake. Swooping petrels and albatrosses kept close company with their limp sails, while seals and whales cavorted in abundance among the surrounding ice. The penguins, threading the pool-like stillness of the surface with cannonball velocity, led the way south, as if toward home. A lucky shot killed a penguin on a floe. When the naturalists opened its stomach, they found pebbles inadvertently swallowed during the assembly of a nest. A quiet excitement spread across the two ships that a major discovery was in reach—not

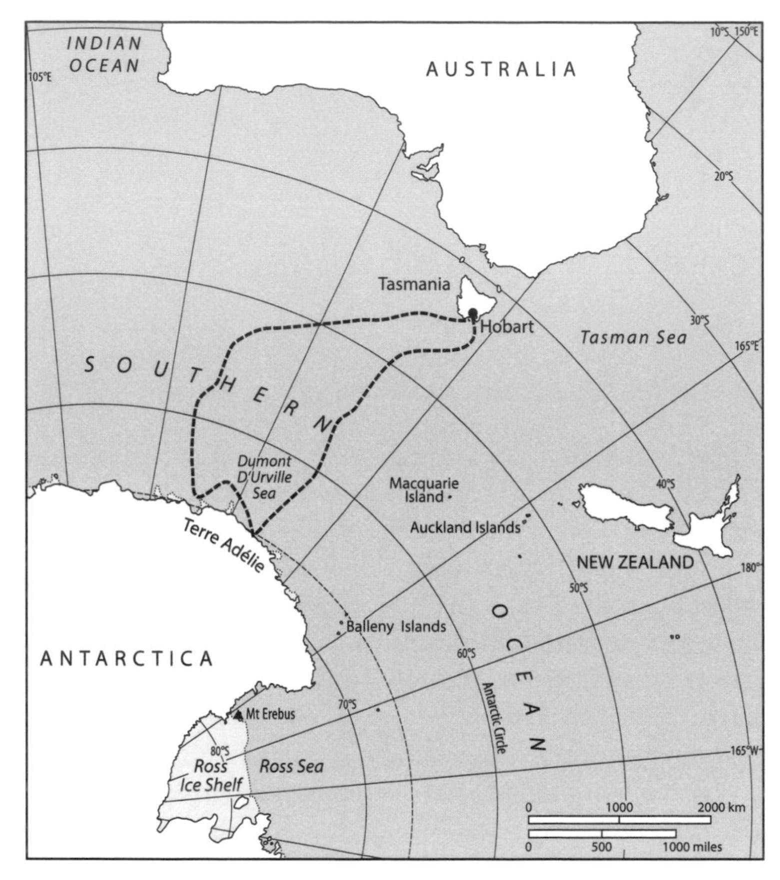

Fig. 9.1. D'Urville's successful second Antarctic campaign, the first to make landfall in East Antarctica.

the South Pole itself, but a great body of land that lay between them and it. *Terra Australis*.

As French hopes grew, so did a skeptical impulse to keep hope in check. On the morning of January 19, a tantalizing black cloud appeared on the horizon to the southwest. To naive eyes, it had all the appearance of land. On the light wind, some thought they caught the discordant, high-pitched chorus of penguins in a mass—nesting grounds, potentially. But an hour later, sure enough, the sky cleared, the fog lifted, and the longed-for "coast" vanished. The officers exchanged knowing looks but climbed back up the masts all the same, spyglasses slung across their backs. Because they were becalmed, D'Urville allowed the men to present a pantomime on deck, titled "Father Antarctica." In the mock ceremony, a man dressed as a penguin announced the founding of a polar academy, Penguinopolis, which D'Urville duly approved. The officers ate penguin for dinner that night. The consensus was it tasted like chicken.

Now the seals, whales, and seabirds had vanished, leaving only the penguins, whose numbers increased markedly. Lines of them, dozens across, advanced in rapid, looping dives toward their unseen destination like a waterborne cavalry. The penguins' constant "kri-kri!" was the only sound in a mute, white world. The division of days lost its meaning as the ships sailed into the permanent dawn and shadowless night of the high polar summer. At some interstice of day and dusk, a dark smudge appeared above the horizon and remained stationary for several hours. None of the young officers claimed to be sure of what he was looking at. The veteran navigator Dumoulin, whose age made him less careful of public mistake, announced the possibility of land (he was, it turned out, the last to be convinced). At eleven, the sun's disk lowered itself gently to the sea, leaving a lurid trail of light above the horizon. From the painterly chiaroscuro emerged a tantalizing contour—a low, rugged coast, if it were real. The men devoured it with their eyes. During the fleeting crepuscular night, all was in suspense, and no one went below.

At two in the morning, the sun reappeared off the port bow. For that historic sunrise, every eye was trained on the suspicious quadrant extending east-southeast to the southwest. The sun's rim illuminated the horizon with ecstatic brilliance, and the next moment a distinct ridge of land appeared boldly before them. Antarctica—the fabled Terra Australis Incognita—announced itself as coastal cliffs of ice terminating abruptly at the dark blue sea. Behind the great, interceding glaciers, glinting plains of ice extended upward, gently but irresistibly, toward an obscure mountainous interior—a land without end. The officers and crew of the *Astrolabe* shouted out loud in triumph.

At the very moment of discovery, the wind failed them completely. So they launched boats to verify the claim. The lucky crews, one each from *Astrolabe* and *Zélée*, stretched out across the unruffled pond. The air was so pure, the light so perfect, the boats were able to keep their mother ships in view throughout the day's discovery campaign. The still distant cliffs of the Antarctic coast offered no suggestion of a landing beach. These were unassailable. But they came among a cluster of islands off the coast where, except on rare days as pristine as this, the surf and wind harassed the tiny line of shore, clearing the ice.

Even so, landing was difficult. One man fell into the freezing sea (he later died of pneumonia). They carried picks and boxes for their precious samples, but the sheer rock at the base of the island was solid granite and yielded nothing. So, even as they felt the frostbite eat their fingers, and their quivering sea legs made them stagger, they scrambled to the island summit. A tribe of penguins observed their ascent with astonishment. The French invaders killed several for specimens, took another alive into the boat, and kicked the rest into the sea. Here a pinnacle of ice had worked the rock into hand-sized fragments, providing the proof of land Commander D'Urville had ordered.

Raising the flag was not a French custom, but the British had made it fashionable, so they brought out the tricolor, plunged the pole into a doughy bank of penguin guano, and shouted "Vive le roi!" Someone had thought to stash a bottle of cham-

Fig. 9.2. "... La decouverte de la Terre Adélie," by Louis Cauvin (1840). D'Urville's Antarctic discoveries became a popular patriotic subject for French artists.—Musée national de la Marine. Creative Commons.

pagne, so they drank it with stiff fingers and blue lips. This French first Antarctic landfall achieved little for natural science, however. Penguins were the sole sign of life on the island. Not a shell, not a weed, not a single lichen could be found on the entire miserable outcrop. They had claimed a colony for France in name only.

Back on board the corvettes, the officers were half dead from cold but giddy with their success. In two years, they had endured enough disappointment and death for a naval lifetime, alongside the usual quota of sheer seagoing boredom. Now, in the space of a single day, they had stumbled into greatness. They had set foot on Terra Australis, the land at the end of the world, and were guaranteed immortality.

Their commander, by contrast, appeared oddly somber. Having planted the French flag on Antarctic soil and, against all odds, won the polar race for his king, D'Urville experienced more relief than triumph. He had escaped humiliation, at least, and so preserved his hard-won fame. But his principal feelings were of a different kind. In keeping with these, and to the silent surprise of his officers, he announced the new continent would be named "Terre Adélie," Adélie Land, and its citizen multitude, the Adélie penguins.

By personalizing the new ice continent, D'Urville rejected both explorer tradition and professional common sense. He named Adélie Land and its signature animal neither after the intrepid men who first described it for science nor for the king, but instead for a woman back in Europe who hated Antarctica with all her heart. Thousands of miles from Paris, in the alternative moral universe of the pole, D'Urville found worldly considerations no longer applied. Since the awful letter at Valparaiso, the voyage of the *Astrolabe* and *Zélée* had contracted, in a sector of his mind, from a national, public expedition to a private passion play between himself and his distant wife, whose forgiveness he yearned for. In a glut of remorse, D'Urville named Terra Australis for his heartbroken partner, whom he had left to face the loss of their son alone. To advertise his feelings as openly as possible, he set aside her formal Christian name, "Adèle," in favor of the diminutive "Adélie"—a loving endearment, a pet name. Thus the great Antarctic landmass, sterile and inhuman, was first named as a gesture of tender apology from an explorer to his abandoned, grieving partner. As for the Adélie penguin—which also mates for life—the name struck an appropriate note of mortality. It spoke the universal chanciness of survival on Glacial Earth, for penguins and humans alike.

Interlude: *Anthropornis nordenskjoldi*

Few dared follow Dumont D'Urville into the Weddell Sea ice for six decades after the Frenchman's near fatal encounter with the ice pack in the southern summer of 1838. Then, in 1901, the Swedish geologist-adventurer Otto Nordenskjöld self-financed an expedition, after his government refused their support. When Nordenskjöld returned home three years later, with nerve-shattering tales of winter on the ice and a cache of amazing Antarctic fossils, the Swedish parliament declared him a national hero and retroactively blessed the whole venture.

In the months before disaster struck his expedition, Nordenskjöld skirted the misty, mountainous coast of the Antarctic Peninsula with the old maps of D'Urville and Ross pinned above his desk. He was mindful of the ghosts of his predecessors as he sailed through D'Urville's majestic Orleans Channel and entered the broad, icy gulf Ross had named for his ships *Erebus* and *Terror*. The towering glaciers reminded him of the Scandinavian Alps, except that this upside-down world allowed him to sail directly to its grand snowy peaks, rather than having to laboriously climb them.

The Adélie penguins were ubiquitous reminders of the preemptive naming rights enjoyed by D'Urville and Ross along the Antarctic Peninsula. And where he discovered unnamed features, Nordenskjöld dutifully honored the polar Victorians. To an unmarked crop of land off the northeast tip of the peninsula he gave the title "D'Urville Island." Then, when circumstances compelled him to make a sledge journey around the foot of Mount Haddington and he discovered it was cut off from the coast, he inscribed "James Ross Island" on his charts. The 1841 Ross voyage, geographically speaking, is more associated with spectacular Antarctic features to the southwest: the Ross Sea, the Ross Ice Shelf, and the twin volcanoes at the end of the world, Mount Erebus and Mount Terror. But Nordenskjöld's tribute to Ross in the Weddell Sea has resonated just as spectacularly in scientific circles. The prehistoric secrets yielded up by the islands collectively known as the James Ross Island Basin have earned it a reputation as the Rosetta Stone of Antarctic paleontology.

As is often the case in the annals of discovery, chance played a leading role. Unaware that his ship, *The Antarctic*, had been crushed and sunk by the Weddell ice pack, Nordenskjöld found himself marooned on the rocky beach of little Seymour Island, under the loom of James Ross Island, for two successive winters. Across a small stretch of water lay the barren, cake-shaped Cockburn Island where, in January 1843, Ross, Francis Crozier, and Joseph Hooker had landed to claim the Antarctic Peninsula for Queen Victoria, and to scour for vegetable life. Hooker came up empty, but the fossils Nordenskjöld stumbled upon during his extended involuntary field trip on neighboring Seymour Island would demand a revolutionary new narrative for the macroevolution of the Southern Hemisphere—its continents, climates, and creatures. In this narrative, desolate Antarctica has been cast in an unlikely role—as a cradle of evolution.

Otto Nordenskjöld, like all visitors to Antarctica, or any well-stocked zoo, was struck by the penguins' comical similarities to humans. The sleek black-and white tailoring of the Adélie, with a rim of white feathers about the eye and fussy gait, evoked a bespectacled older gentleman running late for dinner. In isolation they were a curiosity, but overwhelming in a colony of tens of thousands. At Penguin Point on Seymour Island, the Adélies, their chicks, and their guano covered every inch of ground. They deafened the Swedes with constant shrieking. At every step, the filth splashed over their knees, while the Adélie sentinels attacked them in numbers, head feathers erect and beaks like hammers.

When Nordenskjöld and his men found themselves abandoned on the eve of the winter of 1902, their food supplies dwindling, they returned to the Adélie colony with clubs and guns. The penguins were easy to overpower but hard to kill, sometimes requiring three or four slashes to their long-skulled heads. A day's work brought four hundred Adélie carcasses back to the hut, their killers covered head to foot in the blood and gore of Earth's southernmost nesting bird, a creature adapted over the ages to extremes of environmental change, but not to combat human explorers whose ship has sunk and are running out of food.

Salted penguin meat day after day tasted depressingly like leather, but Nordenskjöld's Seymour Island party did not have the worst of it. The three men sent to find them, led by geologist Gunnar Andersson, were likewise marooned by the loss of *The Antarctic*, but without provisions or equipment for a winter hut. So they killed Adélies of their own and improvised an accommodation out of stones. When Andersson and his men emerged, half alive, in the freezing spring, they began to wander, frostbitten and failing, along the beaches of the James Ross Islands. By a freak chance, Nordenskjöld spotted them in the distance. He thought he was looking at an undiscovered species of lumbering penguin, or was hallucinating. Closer inspection revealed that these polar vagrants were only immersed in the grease, offal, and feathers of the Adélies they had consumed to survive. Face to face with his compatriots, and with no other human beings for thousands of miles, Nordenskjöld still failed to recognize Gunnar Andersson.

The two geologists then revisited the site above the Seymour Island beach where, the summer before, Nordenskjöld had uncovered the stony puzzle pieces of a verdant prehistoric landscape: fossilized trees, leaves, fish, crustaceans, and the hefty limbs of

Fig. 9.3. The Adélie penguin (right), and the Yellow-eyed penguin, native to New Zealand, are remnant species of the penguin's onetime hemispheric dominance.—Dumont D'Urville, *Voyage au Pole Sud et dans l'Océanie sur les Corvettes L'Astrolabe et La Zélée . . . Zoologie* (1842–53). Biodiversity Heritage Library/Smithsonian Libraries.

ancient creatures that had once walked, well fed and confident, on the lifeless terrain that now struggled to support half a dozen desperate human beings. From the bone fragments, they could only guess at the fossilized animals' identity—a wolf perhaps, or a Patagonia-style horse.

Back in Sweden, however, it was established that the Nordenskjöld expedition had made a truly remarkable find: giant penguins. These unusually large ancestors of the Adélies were man-sized, with vicious long beaks for spearing fish and squid. A colony of these monsters would have made short work of the Swedes, no matter how hungry. But the fossilized leaves? Had these been warm-water penguins, beachcombers of a tumid forest? If so, where had this rainforest habitat and its creatures gone?

While these questions hung tantalizingly unanswered, descriptions had to be found for the two inaugural species of giant penguin discovered on the James Ross Island Basin. In this case, the scientific practice of naming new plants and creatures after persons seemed entirely appropriate. The first giant bird was baptized *Anthropornis nordenskjoldi*, for the Swedish expedition leader who had survived a polar winter on Adélie flesh, while the second was named *Palaeeudyptes gunnari*, after the colleague Nordenskjöld had once mistaken for an oversized penguin.

The Adélie penguin—icon of Antarctica—has attracted more research than all other existing penguin species put together. Its giant forebear was tall, graceful, and equipped with a magnificent spear-like beak, little resembling the squat, short-beaked Adélie of modern times. But despite being less than three feet tall and apparently hapless on land, the Adélie penguin dominates the Antarctic ice pack along its eleven-thousand-mile perimeter with a combination of torpedo-like foraging at sea and an efficient domestic breeding culture centered on ice-free pockets of coast. In the crowded colony, male penguins diligently gather stones to make nests and share brooding duties with their partners. When the chick—a silver, sooty ball of fur—emerges, the Adélie parents bring back a dinner of krill and crustaceans from the sea to regurgitate in the chick's keening mouth, but not before making it chase them up and down the beach (reasons unclear). Did giant penguins do the same for their offspring, and did they mate for life, like the faithful Adélies?

Despite the momentousness of Otto Nordenskjöld's fossil discovery, another seventy years passed before paleontologists returned to the James Ross Island Basin. Since the 1970s, however, thousands of prehistoric penguin bones have been recovered from the bare sandstone valley on Seymour Island. Water is almost eight hundred times more dense than air, and early penguins developed durable bones to resist it. For this reason, penguin fossils have a greater presence in the archaeological record than any other prehistoric bird. From the rich repository on Seymour Island, more than a dozen extinct penguin species—including several more giants—have been identified, ten just since 2005. In 2014, discovery was announced of a new giant penguin measuring two meters in length—six and a half feet. Moreover, none of these species can be said to be mere "primitive" antecedents. Each is a fully developed, specialized animal. We are living, in the words of Antarctic fossil hunter Piotr Jadwiszczak, in a "golden epoch" of penguin paleontology.

This long fossil record—tens of millions of years—makes the penguin a unique case history for understanding the relation between species evolution and environmental change in the Antarctic. A puzzling truth emerged from the rash of discoveries on Seymour Island: penguins, unlike the vast majority of birds, boast far fewer species in the present than in the past. Penguins' diversity has shrunk, along with their size. Modern penguins (*Sphenisciformes*) remain circumpolar in their distribution, in keeping with their more massive forebears. Giant penguin bones, of the same species named for Otto Nordenskjöld, have been found on the opposite side of the South Pacific from Seymour Island, in New Zealand and South Australia. But the fossil goldmine on James Ross Island casts our iconic Antarctic *Sphenisciformes*—the Adélie, the gentoo, the chinstrap and the elusive emperor—in a new light.

Given their bird origins, modern penguins boast truly remarkable adaptations to aquatic life. Their wings foreshortened into flippers, while their skeleton uncoiled to allow upright locomotion. The penguin plumage has swapped functions entirely, from aerodynamic loft to heat insulation. Faced with the challenge of drastic climate change, successful penguins changed their diet, too, to take advantage of plentiful reservoirs of planktonic crustaceans beneath the expanding sea ice. As for breeding, Adélie colonies today concentrate exclusively on rocky Antarctic coastline uplifted by deglaciation since the last ice age, so the geographic distribution of current Adélies represents a very recent adaptation.

But by a different measure—deep time—penguins hardly symbolize the triumph of biological adaptation over extreme environments. Rather, through their long species history, dating from the Cretaceous, penguins have suffered wave upon wave of extinction. The penguins alive today constitute a relic miscellany—the thinned-out legacy of a richly diverse population that, in ancient times, enjoyed hemispheric dominance at the pinnacle of the Southern Ocean food chain. Call it the rise and fall of a penguin empire. The sixty-four-million-year question: what triggered the decline?

Penguins have not been the sole fossil dividend of the James Ross Island Basin. By geological accident, the basin offers exposed formations dating in unbroken sequence from the time of the dinosaurs to the Eocene-Oligocene Transition, the momentous transformation in global climate thirty-four million years ago when Hothouse Earth plunged into glacial cold. Seymour Island today is truly bleak, supporting only the odd lichen and moss, and

Fig. 9.4. *Calucechinus antarctica,* a southern beech, belongs to the signature genus (*Nothofagus*) of the high southern latitudes. For Joseph Hooker, the presence of the southern beech in both Australia and South America strongly suggested the existence of ancient land bridges, or continents.—Dumont D'Urville, *Voyage au Pole Sud et dans l'Océanie sur les Corvettes L'Astrolabe et La Zélée . . . Botanique* (1845–53). Biodiversity Heritage Library/Missouri Botanical Garden.

temporary residence for a few seal and penguin species. But evidence for a once-temperate Antarctic climate began with Leg 113 of the Ocean Drilling Program, which drew telltale oxygen isotope data from the floor of the adjacent Weddell Sea. A wealth of fossilized mollusks—the golden keys of paleoclimate—indicated that for millions of years prior to the Big Break, the James Ross Island region belonged to a vast, warm "Weddellian Province," a supercontinental coastline of forests, swamps, sandy beaches, and sunny islands that stretched from South America through Antarctica to Australia. The disposition of tectonic plates played a major role in penguin history, connecting its species evolution to the drift of Southern Hemisphere continents and their changing climates.

For penguins, the Weddellian coast of Gondwana offered a mostly tranquil setting, with open shallow bays protected by barrier islands, and studded with lagoons. Leaf fossils picked up by Otto Nordenskjöld on Seymour Island provided evidence for a coastal forest of conifers, ferns, and beech trees, the last of these being a vital paleoclimatic indicator. The current range of the ubiquitous southern beech (*Nothofagus*) extends across the Antarctic Ocean wastes from Patagonia to Tasmania. But the southern beech is a poor disperser of its seeds: winds cannot carry them, or animals assist via the alimentary canal. The beech tree's only opportunity for hemispheric colonization was therefore the host presence of a continuous landmass, as South America, Antarctica, and Australia were in the last days of Gondwana.

Marsupials also took advantage of the Weddellian land bridge. Cretaceous ancestors of the storied Australian fauna—kangaroos, quolls, koalas, wombats, bandicoots, and Tasmanian devils—scurried west across the *Nothofagus* forest floors of ancient Weddellia. Significant mammalian evolution occurred en route, the secrets of which are buried beneath the polar ice in millions of unmarked marsupial graves. Barren Antarctica, it turns out, is key to the history of charismatic Southern Hemisphere mammals.

Seventy million years ago—in the busy heyday of the Weddellian Province—penguins and albatrosses diverged. For a time, the penguin probably multitasked, both flying and swimming. Then, at the Cretaceous-Tertiary boundary, when a giant meteor crashed into Mexico and turned out the lights on planet Earth, whole niches opened up for species in their scramble to survive. In the chaos, penguins chose the ocean, where they thrived as a new keystone predator. Greater body mass enhanced their superiority in deep-sea diving, while extravagant beaks snared fish unavailable to sharks and other scoop-mouthed carnivores.

Gigantism in the penguin likewise enabled long-distance migration. Amply fueled, *Anthropornis nordenskjoldi* and *Palaeeudyptes gunnari* cruised the entire Weddellian coast, from ancient Chile to New Zealand. The glittering diversity of penguin species also flourished as companions—dividing fish of different sizes amongst themselves. The number of different penguin species living side by side on Seymour Island in their Eocene heyday, fourteen or more, has no modern analogue on Earth. The desert beach where Nordenskjöld's men survived on a single-source diet of Adélies, had been, in post-Cretaceous times, a penguin Manhattan.

Then, through the Eocene-Oligocene Transition, Antarctica became isolated, the climate deteriorated, and penguin populations faced their existential crisis. Ironically, while the Weddellian Province supported an enormous diversity and volume of penguins, it was the final breakup of the Gondwanan supercontinent that spawned the evolution of penguins as we know them today, which adapted to the cooling water temperatures produced by the Drake Passage and modern circulation patterns of the Southern Ocean. For the many ghost lineages of extinct penguins, however, their Goldilocks climatic tastes—not too hot, not too cold—brought them to grief in the new glacial regime.

The decline of the giant penguins and the birth of modern penguin species broadly coincided about forty million years ago.

Unable to compete with the newly ascendant whales of the Southern Ocean, the man-sized penguins named for Otto Nordenskjöld and Gunnar Andersson died out, while the emperor penguin diverged with the new *Pygoscelis* lineage—our Adélies, chinstraps, and gentoos—to form a sleeker, smaller brood of cold-water hunters, adaptable to the expanding ice. As the planet cooled across the EOT and beyond, more penguin species succumbed. It was a case of colonize or die. Penguins migrated northward out of the expanding glacial zone, following the warmer waters to which they were accustomed. Most failed, but the survivors established the miniature extratropical populations we know today, from southern Australia to the Galapagos. To our scrambled perceptions, these fair-weather species constitute the penguin anomaly, when the opposite is true. The Adélies and their Antarctic compeers are the exception that proves the rule. Historically, most penguins, like most humans, don't like the cold.

Pack-ice penguins such as the Adélie forage for long periods in waters well below their core body temperature. Whereas, in temperate and tropical birds, blood is supplied from the body to the wingtip via a single artery, the modern Antarctic penguin sports multiple arteries branching into veins along the wings. Warm blood from the penguin core circulates continually outward, moderating the incoming supply from the wing to maintain core temperature. Thanks to this complex heat-retention structure, the differential between a penguin's body and wingtip can be as much as 30°C degrees.

The irony of the modern penguin wing's vascular heat exchange is that it evolved on Hothouse Earth fifty million years ago, *not* in response to late Eocene global cooling and Antarctic ice. Rather, wing restructuring belonged to a suite of skeletal modifications that rebuilt the penguin for success as a dominant underwater flyer in the new marine order. These included thickened plumage, a gigantic body for deep diving, dense bones for buoyancy, and wings optimized for drag reduction. The extraordinary hydrofoil wing, in particular, equipped with a heat-retentive capacity unique among birds, preserved energy, extended feeding range, and proved the key to penguin survival on Glacial Earth.

When the first ice came to Antarctica, and temperate-adapted species, including a majority of penguins, vanished in droves, an insulated, wing-powered penguin subset discovered its new niche beneath the ice and on the frozen, barren coasts. To the French explorers of 1840, the two-toned flipper-wings of the Adélies were signature appendages of a ridiculous creature, useful only for target practice or to boil into soup. But that stiff, truncated, arm-like limb told an unrecognized story of resilience. It meant it was the Adélie penguin that saluted the newcomers from their icy ramparts, millions strong, and not some other creature—or nothing at all.

‡ 10 ‡

Wilkes Discovers a Continent

When, back in 1836, the United States Congress had at last approved funding for the round-the-world expedition, the Navy commissioned an exploration fleet on the model of the British polar vessels *Erebus* and *Terror*. The result was an expensive fiasco. The newly minted ice-strengthened ships—a frigate, two brigs, and a schooner—were so overburdened by their massive timbers they could hardly be boarded safely, let alone trusted to the open ocean. So Wilkes had to make do with a last-minute, hodgepodge flotilla, without structural insurance against the ice.

In Sydney, in December 1839, Wilkes conferred with his second-in-command William Hudson over their continuing concern for the *Peacock*. She was built of strong oak and sailed with a queenly mien, but her gunports warped and leaked water, and a recent near-shipwreck in the Persian Gulf—two days stuck on a reef—had subtly opened her seams. The criminal unfitness of the American squadron had already been exposed by the loss of the *Sea Gull*. Now the remainder of the fleet was to plunge into the Antarctic maelstrom once again. The captains agreed the *Peacock*, despite refitting in Rio, was not fit for polar seas, but they likewise agreed that she must go regardless. National honor was at stake.

The American squadron sat moored in Sydney Harbor adjacent to Fort Macquarie, the location today of the resplendent opera house, where it was the opinion of locals snooping around the American ships that they were no better than floating coffins. No reinforced hulls to buffer collisions with the pack; no prow saws to cut through the ice; no watertight compartments for buoyancy; no modern heating system; and only limited stowage for fuel. If the ships were to become trapped in the ice, the coal would run out midwinter, and the food soon after. In short, the Americans had quantity but not quality—plenty of ships, but none as polar-ready as D'Urville's *Astrolabe* and *Zélée*, with which, as New Year 1840, dawned, they were due to compete in a head-to-head sailing duel to the South Pole, winner take all.

Worse even than the state of the ships, however, was the unstable condition of their commander. At least, such was the opinion of surgeon Edward Gilchrist, sitting despondently on his bunk, confined to quarters on the *Vincennes*. He, like the rest, had embarked aboard the flagship with highest hopes. Since Rio, however, his enthusiasm had gradually given way to creeping dread. Gilchrist saw that Wilkes overworked himself to the point of collapse. The captain rarely slept five hours out of twenty-four and was in a constant lather. He failed the first test of command—by refusing to delegate responsibility. In addition to controlling every detail of the squadron's organization, he insisted on supervising the scientific work, including tedious meteorological record-keeping and magnetic observations. The more capable the officer, the more Wilkes distrusted him, as if he feared his own abilities might be outshone. And among the thousand duties he kept to himself, he was unable to prioritize. One moment he was on deck with his trumpet, yelling "all hands" for no reason. The next found him in the infested midshipmen's mess, squishing spiders like a man possessed.

Gilchrist's medical experience told him that the man must break. Sure enough, onshore in Rio, Wilkes had collapsed into unconsciousness after an evening bath. With worried officers gathered outside the door, Gilchrist pronounced the commander's condition "very serious." He prescribed drugs and rest, but a recovering Wilkes waved him away. The next morning the commander reemerged, as unhinged and frantic as ever.

Wilkes's only relief from constant headaches came in tormenting the officers, whom he accused of conspiring against him. He mocked the best of them on deck in front of the ship's company, threatened court-martial, and reassigned them from ship to ship so they could not settle. The men suffered equally. Drunken hands got twenty-four lashes, twice the number permitted by regulations. Deserters got more, without the required court-martial.

Tyrannical antics among ships' captains were hardly unknown at sea, and might have been endured. But doubts about Wilkes's seamanship hung like a cloud over the US Ex. Ex., because mismanagement of the ships threatened the life of every last man. Wilkes had a particular fetish for keeping the squadron in close company, a symptom of his poisonous distrust. One quiet morning off Tahiti, he became suddenly enraged with the easy progress of *Flying Fish*—whose commander he particularly hated—and ordered the schooner to head up into the wind and wait. Lieutenant Pinkney did not hear the inexplicable command, until he looked up to see Wilkes running along the deck of the *Vincennes* with his trumpet, screaming at him to heave to. Now panicked himself, Pinkney abruptly obeyed, even though the maneuver would bring tiny *Flying Fish* under the flagship's bows. Only a quick order from an officer on the *Vincennes* to trim sail saved the schooner from destruction.

Then there was the embarrassing debacle at Pago Pago, when the *Vincennes* "missed stays" and drifted listlessly toward the rocks at the mouth of the Samoa harbor. The officers looked to their commander for steadying orders, only to find Wilkes had disappeared. He was hiding by the gangway, face in his hands, while the local English pilot averted wreck for the *Vincennes* at the last possible moment. Gilchrist's assistant reported to him that Wilkes had utterly broken down in the crisis, exhibiting "confusion and alarm," and was "incompetent for some time to his duties."

The mutinous idea that the commander was incompetent caused a revolution in Edward Gilchrist's attitude toward the glorious exploring expedition. Wilkes must be relieved, or the whole enterprise abandoned. First, he quietly raised the issue of the commander's fitness among the officers. But Wilkes had ears everywhere on the *Vincennes* and confined the surgeon as punishment. The second campaign to the South Pole loomed—a suicide mission. In desperation, Gilchrist tried to save himself. He wrote a deliberately provocative letter to Wilkes questioning his command, concluding with a request he be released from the expedition in Sydney. But Wilkes had already lost several of his medical staff and refused the request. So, as the American squadron prepared to sail south in January 1840, Gilchrist found himself once more imprisoned in his berth, at the mercy of a man he considered an incompetent maniac, on a ship destined for a graveyard of ice and whatever fatal shores lay beyond. As one American midshipman confided to his journal, probably not a single man of the US Ex. Ex. believed the responsibility for their survival could sanely be entrusted to "the hero of Pago Pago."

With Antarctica once again in prospect, Wilkes faced an epidemic of desertion. Not even a reward of one hundred fifty dollars a man, and the resourcefulness of the Sydney police, could locate his crew disinclined for the pole. "It will be a great wonder," wrote another sailor in his journal, "if we return from the South." It was fortunate, perhaps, that he and his messmates did not appreciate the worst of it—that they *underestimated* the danger. Rounding south of the Australian mainland, the green coast of Tasmania disappearing astern, the Stormy Petrel and his ships were making a direct course for the most gale-prone stretch of coast on Earth.

The young officers of the American expedition enjoyed a complacent faith in their invincibility. Instead of shipwreck in the ice, their concern with Commander Wilkes focused on what they perceived as his despicable stratagems to hog the glory of polar discovery for himself. Perhaps their fears were well founded, or perhaps Wilkes had so thoroughly alienated his officers by early 1840 that they put the worst possible construction on all his actions. In either case, the squadron journeyed south under the baffling instructions to keep in close order, when the goal of discovery would obviously have been served by spreading out. Wilkes also ordered an early rendezvous at Macquarie Island—where noisy nesting seabirds blanketed every last inch of rock—an appointment the *Vincennes* herself failed to keep. Wilkes's wicked design, his enemy compatriots believed, was to give himself a head start. The race to the pole had now become a race within a race. The British and French were Lord-knows-where, leaving the feuding Americans to wage the battle for glory among themselves.

Four American ships crossed the Antarctic Circle in January 1840. The three-masted flagship *Vincennes* dominated the group. Next came the *Peacock*—like the *Vincennes* designed for warfare, not Southern Ocean exploration. Then the smaller, lightly built brig *Porpoise*. Finally, at the rear, came the least polar-ready vessel of the four. *Flying Fish* was ideal for surveying small islands in the Pacific, though she had also performed the best of the fleet during the previous summer's abortive

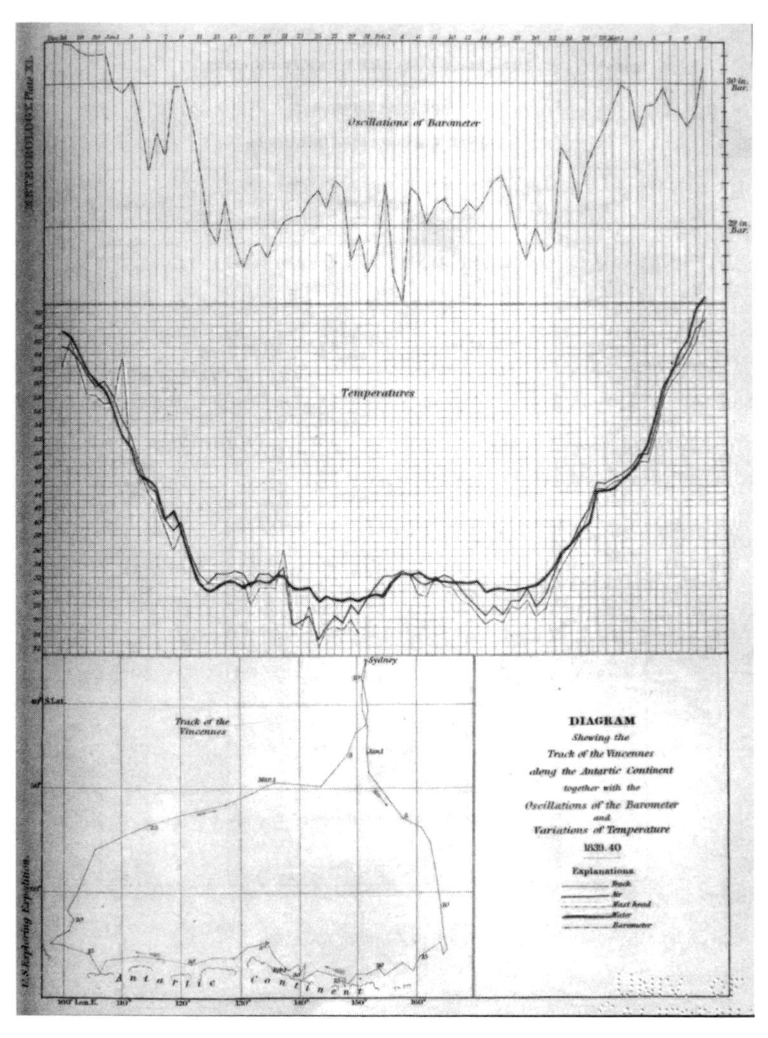

Fig. 10.1. Wilkes's map of the *Vincennes*'s route of discovery along the East Antarctic coast in early 1840 is combined with barometric and temperature data for the same period. Note the low barometric readings during the great storm that almost sank the American fleet.—Charles Wilkes, *United States Exploring Expedition*, vol. 11, *Meteorology* (Philadelphia: C. Sherman, 1851). Biodiversity Heritage Library/ University of California Libraries.

assault on the pole, rivaling Cook's *ne plus ultra* under the ambitious captaincy of William Walker. But the sheer terror of navigating the pack ice in a tiny schooner loomed larger in the memories of her crew than any geographical prize. Half deserted in Sydney, leaving *Flying Fish* undermanned with a ragtag, dispirited company. For all four nervous crews, a further blow came on opening the winter clothing stores, courtesy of a thrifty US government contract. Their boots leaked, while their cloth pea jackets were next to useless in the polar wind.

The *Peacock* began well. Her course soon took her beyond the busy southern fishery, which more than a hundred whalers patrolled, most of them American. She passed her first icebergs in mild weather. The sun glinting off the emerald ice dazzled their eyes and imaginations. Midshipman William Reynolds, recently expelled from the *Vincennes* for disrespecting the commander, was sure of the *Peacock* beating Cook and Weddell and passing 75° south into the polar realm "that was neither earth nor sea." Then they met with the ice pack and their dreams evaporated. A wasteland of ice stretched across the horizon, blocking their passage. On reaching the ice pack, their albatross companions deserted the ship, taking luck with them. But the smaller petrels remained, perching sometimes on the ice or hectically skimming the water for food.

With little for the *Peacock* to do but cruise along the rim, Reynolds spent the morning of January 17 at the masthead with his friend Midshipman Henry Eld from Connecticut. From there they observed, quite clearly, a range of mountains beyond the ice—solemn peaks shrouded in mist. They called for a spyglass to confirm the sighting, then hurried to inform Captain Hudson. For reasons he was never afterward able to explain, Hudson did not come on deck to see the longed-for Antarctic land for himself. Perhaps he had slept badly in the permanent light, or was too comfortable by his stove to bother yet again rugging up against the damn cold. Either way, he was unimpressed by the breathless reports of his midshipmen, and their discovery went unrecorded in the *Peacock*'s log. Later, to their dismay, they learned that Dumont D'Urville had logged his sighting of the Adélie Coast on the afternoon of that same day. Official discovery of a continent millions of years old had come down to a few hours, and one captain's inexplicable mental lapse. Wilkes routinely bewildered his officers with his rages and obsessions, but it was William Hudson—the reliable Navy professional—who ended up baffling all their polar ambitions, and history itself.

It required all Hudson's seafaring skill, however, to bring the *Peacock* back from the ice. Disoriented in the pack, she rode backward onto a jagged floe, ripping her rudder off the sternpost. Stranded thousands of miles from Sydney and surrounded by ice, the *Peacock* could not steer. She soon crashed into an iceberg, sending splintered spars from the upper masts crashing, and men sprawling onto the deck. The stern boat had been smashed into a thousand pieces. A second collision would sink them. Looking up, William Reynolds saw a black "funeral cloud" advancing toward them and "gave up all hope of life." But in this land of mirages, the squall vanished, and the *Peacock* slowly limped north toward safety. On January 27, they met with the Adélie Coast storm that was to be the fleet's immortal contribution to polar meteorology. Three weeks later they reached Sydney, grateful for their lives, and with only a conspicuous blank in the log to signify their lost chance at fame.

Survival not fame, meanwhile, was the principal concern aboard the long-suffering *Flying Fish*. The last-minute recruits from the Sydney docks proved useless, forcing the officers to join the crew to work the ship. As early as New Year's Day, they lost their jib in rough weather. Wilkes, observing this calamity

Fig. 10.2. The *Peacock*'s collision with an iceberg during the 1840 polar campaign was a near catastrophe for the American expedition.—Charles Wilkes, *Narrative of the United States Exploring Expedition* (Philadelphia: Lea and Blanchard, 1845). Rare Book and Manuscript Library, University of Illinois at Urbana-Champaign.

from the deck of the *Vincennes*, signaled "make sail" and cruised away, which the incredulous crew of the *Flying Fish* could only interpret as a sick joke. Quick repairs enabled the schooner to reach as far south as the ice pack, but the three-day storm in late January swamped the decks and opened her seams. Officers and crew manned the pumps in their near-frozen state, and soon the last dry square foot aboard, the stove in the officers' cabin, was inundated. Their clothes and bedding soaked, the men slipped one by one into hypothermic despair. In their extremity, they composed a letter to Lieutenant Pinkney (sitting a few feet away), expressing their refusal to continue on a course that "must soon terminate in DEATH." A very reasonable threat of mutiny. The next day, February 5, Pinkney turned back north.

The *Porpoise* had the oddest cruise of the four American ships. On reaching the pack, they tacked first north then west, taking them away from the land that Reynolds and Eld had seen aboard the *Peacock*. They then hauled off farther north still to sit out the terrible gale, a course that, by pure happenstance, enabled their encounter with the French ships *Astrolabe* and *Zélée*, returning from their own polar adventures that season. In those vast wastes of unexplored ocean, the Americans' shock at meeting another ship was profound, like Coleridge's Ancient Mariner spotting the phantom ship sailing without assistance of the wind and manned by ghouls. The commander of the *Porpoise*, Cadwallader Ringgold, first assumed the strangers appearing out from the mist were *Erebus* and *Terror*, and prepared to salute the celebrated discoverer of the North Magnetic Pole. But when the French flag appeared at the mast, Ringgold realized it was D'Urville, who abruptly made sail and sped away at his approach. Ringgold was dumbfounded, then suspicious. What had D'Urville found that he did not wish to share? This surreal meeting with their French rival was the only story the *Porpoise* had to show for six weeks in the ice.

The last hope for American glory, then, rested with the *Vincennes*. In the opinion of the resentful officers of the squadron, this was just as Wilkes had planned it all along. On January 3, the flagship entered the land of mist and snow, and soon lost sight of the *Peacock*. A week later, they reached the pack—a flat, undulating plain of ice dotted with squarish, mountainous bergs. The sea was calm, as if asleep. Their voices reverberated magnificently against the massive ice walls. Otherwise, only the low rustling of the water at work on the pack broke the silence. At first, they observed the ice coating on the rigging as a thing of wonder rather than fear. Birds appeared at the *Vincennes*'s stern and cruised along her bow, inspiring hope that land was in the offing. They flew in and around icebergs shaped like arches and crowned with turrets, like fairy birds in a white-world castle. A new species—a perfect, snow-white petrel—seemed like an emissary from the pole itself. They shot and stuffed it for their collection.

By January 23, all aboard the *Vincennes* were convinced they saw the distinct outline of land—too real, too permanent for an illusion. The appearance of the coast was clearest on the morning of January 28 when Wilkes, navigating cautiously southward through an opening in the pack, began to seek out a place to land and raise the Stars and Stripes. The glory he was convinced was his due lay within tantalizing reach; it would never be closer.

Then a thick fog descended, the wind sprang up, and Wilkes's thoughts turned reluctantly to the perilous situation of the *Vincennes* trapped in the pack in a gale. Their retreat was littered with icebergs, which seemed to close in just as the storm intensified. The dense blizzard snow flew horizontal to the deck, choking them, and reduced visibility to a ship's length. Every wave that crashed across the bows left a fresh sheen of ice across the deck and rigging. One man slipped and broke his ribs. Another became stuck on the yardarm hauling in a sail. By the time he was spotted and dragged down, he was frozen stiff and barely alive.

Below deck, Edward Gilchrist tried not to contemplate the accumulating horror of their situation. He had been released from confinement to tend to the sick and injured, who were stumbling below in a steady stream. The men's skin was erupting in boils from exposure to the cold. Light scratches became bleeding, ulcerous wounds. Cases of frostbite, fever, broken bones, and sheer exhaustion crowded the sick bay. Many others in similar condition, he knew, remained on deck, not wanting to abandon their posts in the ship's hour of need. Then word came that Wilkes had turned the ship south, *back* into the screaming void. After five days in an upended hell—groping through the lurching dimness from cot to cot—the merely sick were groaning, while the rest had gone inhumanly quiet: men who were resigned to death and who had begun to die. The doctors had had enough. They passed a letter to their mad commander, demanding he sail north immediately to warmer latitudes, out of the storm. With the formality of a quadrille, Wilkes passed the surgeons' request to his own officers, who then submitted their own letters, agreeing to abandon the attempt on the pole, to save the men.

Having taken the trouble to consult his officers and staff, Wilkes ignored them. It was fitting that the man they called Stormy Petrel brought the heaving *Vincennes* with an unwilling crew across the storm-tossed Antarctic Circumpolar Trough to within a half mile of the dark rocks of Adélie Land. He then sailed fifteen hundred miles along the high, plainly visible coast. Before the *Vincennes*'s voyage, "Antarctica" was reckoned to be, at most, some scattered islands, or a frozen archipelago opening to a great polar sea. In January 1840, Charles Wilkes redrew this map, and the world's entire conception of Terra Australia Incognita. At 66° south, in the midst of a howling gale—in a place he named Piner Bay—he pronounced this new, unbroken line of coast the visible extremity of an "Antarctic Continent."

With his epic, fifteen-hundred-mile coastal navigation of Antarctica, Wilkes had achieved what neither of his celebrated French and British rivals would, under the worst possible circumstances and without the skills that gave him a right to succeed. It didn't help that his compatriots scarcely thought him capable of what he had done, let alone deserving. When he invited the *Vincennes*'s officers to his cabin to celebrate their grand

Fig. 10.3. The American flagship, USS *Vincennes,* in Disappointment Bay, so named for the explorers' being denied landing along the heavily glaciated East Antarctic coast. Disappointment Bay was also the scene of a nasty confrontation between Commander Wilkes and one of his officers, an all-too-common occurrence on the US Exploring Expedition.—after Charles Wilkes (1843). Peabody Essex Museum, Salem, MA. Photo: Mark Sexton.

discovery with champagne, some dared to call him "d——d lucky." So the captain banned them. Bloodymindedness had earned him his glory, and it ensured he made only enemies in the process. Even as Charles Wilkes sailed exultantly northward—Discoverer of Antarctica!—forces were at work that would strip him of all credit for that lustrous achievement.

When the great storm finally abated, the *Vincennes* cruised back north toward Australia, into warm and pleasant seas. With the crisis over, Wilkes finally collapsed. He could not sit up in his bed or reach for a cup without a fit of tremors. On the night of February 22 came a welcome respite when the people of the *Vincennes* were treated to an exhibition of the aurora australis. A bright orange nucleus appeared around the clouds above the mist, then spread across the sky forming a luminous halo. Shooting rays lit the dark clouds from beneath, setting them in bright relief against the sky, like pasteboard clouds on an opera stage. Electric rays darted to the horizon, gathered themselves in a ball, then flung and folded upward in an ecstasy of light.

Wilkes lay on his back on deck to give himself the best view of this marvel of nature. The polar auroras are the work of the sun, dispatching charged particles at light speed toward the atmosphere. On contact with the Earth's magnetic field, the solar energy discharges itself toward the poles, north and south. To Charles Wilkes, in his time of triumph, the aurora appeared like a heavenly salute. His head resting on the hard deck, he could almost see himself in the electrified sky—the Stormy Petrel riding high on a solar wind.

Interlude: The Hurricane Coast

South of the storm track that, in late April 1839, sank the schooner *Sea Gull*, the Antarctic Circumpolar Trough extends from 60° south to the icebound coast of the Antarctic continent. At the coast, however, the storm zone abruptly terminates. The forbidding polar vortex—cold, bone-dry air off the Antarctic cap—functions like an atmospheric drawbridge, securing the ice fortress as a climate realm unto itself. Incredibly, however, the coastal winds are even stronger than those of the Circumpolar Trough or the Roaring Forties, as the captain and crew of the USS *Vincennes* discovered in their historic campaign of January 1840.

Seven decades passed before another vessel ventured in the wake of the US Exploring Expedition and its French rivals to the East Antarctic coast. In January 1912, an Australian expedition led by Douglas Mawson became the first team of scientists to establish a station in the polar wilderness of the Adélie Coast. From the start, they battled hurricane-force winds and raging seas. Once established on shore with their dogs and provisions, their first act was to toast the memories of D'Urville and Wilkes, who had blazed this trail, unimaginably, under sail in wooden ships. Mawson named the expedition's first Antarctic-born huskie puppy, "Admiral D'Urville." No less generous to the memory of his American predecessor, Mawson designated the entire territory "Wilkes Land": it remains the largest area on Earth named for a single individual.

After a month penned in by continuous gales on the Adélie Coast, it dawned on Mawson that his best contribution to science might lie in study of its unique meteorology. The relentless hurricanes and whipping snow thwarted most other research. The men spent weeks confined to the freezing hut, devising safeguards against being blown away. Outside, the wind oscillated from a dull howl to a deafening scream. Any man not wearing crampons on his boots was sent spinning across the ice until he collided with a boulder or was buried in drift. To avoid being instantly flattened, the men designed a polar style of walking, leaning at 45° and allowing the wind to prop them up.

The expedition meteorologist, Cecil Madigan, returned from daily check of his instruments with a mask of ice across his face—a crystallization of human breath and the blizzard-blown snow. One memorable morning, his cabin mates watched with morbid curiosity as Madigan tried to peel his cheek from his face, thinking it was ice. The rare occasions the winds abated, their ears still continued to ring with a high-pitched drone. They couldn't speak without shouting. In this inverted, polar hell, silence roused them from sleep, hungering for the "incessant, seething roar" of the katabatic winds.

The Greek *katabasis* means "descending." The persistent, hurricane-strength gales along the Antarctic coast begin as parcels of sinking cold air that stream from the high plateau of the interior. Reaching the steep mountain slopes near the coast, they abruptly intensify before barreling out to sea. But why so extreme at Adélie Land? Mountainous coasts are found the world over. Even within Antarctica itself, the winds along the Adélie Coast are 70 percent stronger than elsewhere, Earth's nearest analogy to the permanent, sirocco-strength gales of Venus. For polar meteorologists—Mawson's experience vivid in their minds—the extraordinary microclimate of Adélie Land long remained an enigma.

The sled-driven land excursions conducted by the Australians had discovered a peculiar feature of the Adélie winds. The katabatic funnel extended at least sixty miles inland to the south and two hundred miles to the west, but only a short distance east.

On one exploratory trek, Mawson lost one man and his dogs down a crevasse, and another to toxic dog meat (the glaciers Ninnis and Mertz bear the men's names). With the soles of his feet virtually peeled off, Mawson continued alone, crawling the last hundred miles through a howling "void . . . grisly, fierce, and appalling." Though the competition is stiff, Mawson's ordeal in Adélie Land must stand as the greatest solo tale of polar survival. Regardless, the Australian meteorological community refused to honor his hard-won data. Mawson's instruments were requisitioned, tested in a wind tunnel in Melbourne, and found to be "faulty." Madigan was directed to revise the weather records downward by as much as 20 percent, so not to compromise the reputation of Australian polar science with his outlandish numbers.

Three decades later, however, in 1951, a French team vindicated Mawson's data. The annual mean wind speed on the Adélie Coast amounted to "strong gale force" on the Beaufort scale, while the mean for the wild month of March 1951—twenty-nine meters per second—exceeded by far any known from other stormy places on Earth. The French endured hurricane-force winds every third day, blowing with eerie steadiness from the southern interior. Truly end-of-the-world winds. The annual volume of snow transported by these blizzards—seventy million tons for each kilometer of coast—exceeds the loss of ice through the calving and melting of glaciers. But while the French data restored Madigan's reputation, the larger mystery of Adélie Land weather persisted. In a summary essay from 1972, titled "Land of Storms," Australian meteorologist Frederick Loewe issued a challenge to the polar science community: the most spectacular, dominant feature of Antarctic climate—its terrifying coastal gales—remained unexplained.

In the late 1970s, radio-sounding technology, employed on transcontinental flyovers, enabled the topography of the Antarctic interior to be mapped for the first time. This opened the door for a young American meteorologist, Thomas Parish, to tackle the question of katabatic winds. The maps revealed Antarctica as a great carapace of ice, its vast, lifeless interior a gently sloping terrain in the character of the American Great Plains. With average elevation over two kilometers, no weather system from the ocean could penetrate the mountainous Antarctic, leaving the white continent to manufacture its climate according to its own rules.

In the world at large, pressure systems of the upper atmosphere largely determine the weather we experience at ground level. But in Antarctica, topography drives climate. The first close-grained maps captured the wavelike undulations of ice through which the light interior winds channel and converge, gradually gathering force. This continental ice plateau, Parish reasoned, serves as both engine and fuel for the weird katabatic winds at the coast. Unlike standard conditions on Earth, where temperature decreases with altitude, the Antarctic surface promotes thermal inversion, with warm air displaced upward (thermal inversion refrigerates the Antarctic ice cap by trapping cold air at the surface). This vast reservoir of cold continental air then sinks, pouring radially outward along the surface ridges of ice, which act like finger-drawn hieroglyphs of the winds' course.

Nowhere else on Earth does a single meteorological element define a continent's climate as katabatic winds define Antarctica. Approaching the coast, the gravitational pull of the slope escalates the stiff breeze, by increments, to a shrieking gale. This freezing wind then enwraps the warm frontal currents of the Southern Ocean, rising and decelerating rapidly, as an albatross banks upward from the wave. Creating a wind model based on the new maps, Parish determined that the topographical conditions for katabasis applied most dramatically on the Adélie Coast where, like loose strands of a rope knotted together, a string of ice ridges converged, by geological happenstance, on a single, narrow coastal zone: the "land of storms." It was to this precise point, on the entire Antarctic coast, that Charles Wilkes unwittingly led his motley flotilla in January 1840.

Parish published his solution to the longstanding puzzle of the Adélie winds in *Nature* in 1987, but the wild katabatic phenomenon still tugged at him. How did these land-driven winds, of apocalyptic force, interact with the greater synoptic features of Antarctic climate, such as the polar vortex and coastal cyclones? Revisiting the subject in the first decade of the twenty-first century, Parish could include the rich data returns of the modern Antarctic Weather Station Program, digested within next-generation Antarctic climate models. He nevertheless found himself drawn back to his original source—the data of the 1912–13 Mawson expedition. He reanalyzed Madigan's readings, seeking answers to why the battered strip of coast that plays host to the most persistently brutal winds on Earth is likewise the scene of such frequent, violent storms offshore. For original data on this score, however, the Mawson expedition must share credit. While the Australians could claim to be the meteorological pioneers of katabatic winds, it was the Stormy Petrel himself, Charles Wilkes, who left the first synoptic account of an Adélie Coast cyclone—a true Antarctic widowmaker.

Back in 1831, the pioneer American meteorologist William Redfield had defined wind as "air in motion," a storm as a "violent wind," and a hurricane as "a wind of the most extraordinary violence." He understood that the relation between the wind and ocean currents was both analogical and material. *Analogical* because, in its physical dynamics, wind behaves like water in a current (Mawson described the katabatic winds of the Adélie Coast as "a river . . . a torrent of air"). *Material* because the hemispheric circuit of westerlies and trade winds physically drives ocean currents on Earth. The sovereign kingdoms of air, land, and sea are part, in fact, of one interconnected Earth system.

Redfield observed, furthermore—in his paper bursting with genius intuitions—that "where winds moving in different directions are brought to bear upon each other . . . the violent rotative effects naturally follow." Without knowledge of polar climatology, he nonetheless described precisely the interaction of the fierce katabatic winds of Adélie Land with the eastward-bearing cyclonic storms native to the coast. In the southern summer of 1840, Redfield's compatriots in the US Ex. Ex. sailed—without charts or forecasts—into a unique weather province driven by mutually reinforcing extremes. On the Adélie Coast, cold land-sourced katabatic winds enhanced warm-air offshore frontal systems, while those ensuing cyclones in turn supercharged the winds.

Safely back in Washington in 1843, Wilkes pored over the ships' logs for the 1840 Antarctic campaign, piecing together the great storm's progress. The waters the American squadron sailed into, near 150° east, off the Adélie Coast, have since been identified as the deepest well of cyclonic storm formation in East Antarctica. On January 28, 1840, the *Porpoise* was the first of the American squadron to be hit. Situated at the western edge of the storm, she was pummeled by gale-force winds from the east, then south. A mere sixty miles away, to the south-east, the *Vincennes* then crossed into the path of the same easterly gale, while the unlucky crew of *Flying Fish*, two hundred sixty miles to the east, were spared until the following day for their life-and-death battle with a blizzard out of the *northeast.* Lastly, the *Peacock*, another one hundred forty miles to the northeast, and badly winged from her accident in the pack, bore the worst of it for longest, and without a rudder.

Wilkes calculated the speed of the churning system at twenty miles an hour. The still center of the storm, he reckoned, must have passed between the four ships, with the *Peacock* at its northern rim, and the *Flying Fish*, *Vincennes*, and *Porpoise* to the south. Slow-moving, vortical, and monstrous, it was a classic Antarctic storm, as modeled by William Redfield. The Americans' firsthand experience of the deadly system—combining real-time data with the latest synoptic model—was as valuable as any exotic specimen the US Ex. Ex. brought home to the Smithsonian. As the forerunner of Mawson, Madigan, and Thomas Parish, Charles Wilkes—discoverer of the Antarctic continent—warrants an additional distinction. He was the first to upload Antarctic extreme weather data into the new meteorology's "vast machine."

⚓ 11 ⚓
Message in a Bottle

On the arrival of the *Erebus* and *Terror* in Hobart in August 1841, James Ross assured his fiancée in a letter that his quest for the South Magnetic Pole was on track. Although his rivals Wilkes and D'Urville had taken magnetic readings at record-high latitudes the previous summer and made speculations on the position of the South Magnetic Pole, "nothing short of reaching the pole itself will satisfy the demand of science." The door, therefore, stood open to him. By mid-January, he told Anne, "I hope to discover something worthy of being distinguished by the name I love."

In reality, Ross was less certain of his prospects. Was the magnetic pole located on land or sea? Was it approachable from the north or east? Did it lie behind some insuperable barrier? While waiting for the summer exploration season to commence, Ross embraced his duties as a research magnetician as relief from anxiety over the French and American expeditions and the mysterious situation of the pole. To fulfill the quest for global magnetic data, the *Erebus* and *Terror* had been equipped with state-of-the-art dip circles designed to operate on the decks of iron-girded ships in rough seas. When the ships were at anchor, flat packs appeared from the holds, and portable magnetic observatories sprang up, like artificial trees, on remote beaches in the Indian and Southern Oceans.

Now a largely forgotten episode in the history of science, the international mania for terrestrial magnetism fueled the nineteenth century's first "big data" project and provided important momentum for the Antarctic discovery voyages. The Magnetic Crusade, as it was called, consumed vast intellectual and financial resources over decades, and required coordination of teams of researchers across the globe. As such, it provided a template for the practice of modern science: professionalized, institutional, and tightly enmeshed with social and military agendas.

Throughout the four years of his southern voyage, Ross kept up a detailed magnetic correspondence with Edward Sabine, his old shipmate from the Arctic and British standard-bearer of the Magnetic Crusade. A Dublin-born artillery officer, Sabine had returned from the Napoleonic Wars with a distaste for army life and vague scientific ambitions. At a loose end in 1815, he, like most educated Britons that year, devoured the first published English translation of Alexander von Humboldt's celebrated *Travels* in South America. Humboldt was the preeminent scientific entrepreneur of the age and the first champion of international magnetic research. He traversed the Andes taking readings, which offered tantalizing evidence that magnetic intensity was not constant across Earth, but mysteriously increased with latitude. In magnetism, for Humboldt, lay the secrets of Earth, sun, and sky.

For his enraptured convert, Edward Sabine, the baffling inconsistencies of magnetic forces at play across Earth, punctuated by the spectacular magnetic "storms" of the polar auroras, encapsulated the pure romance of science. Data gathering at this scale demanded extreme travel, heroic fortitude, and painstaking observations in remote places. Humboldt's Magnetic Crusade gave instant purpose to Sabine's postwar existence and shaped a fifty-year career at the pinnacle of the British research establishment. Still attached to the army, he continued to earn promotions for his magnetic charts, scientific papers, and staggering reams of data. He acceded to the presidency of the Royal Society in 1861—as the very model of the modern major-general.

At the outset of Sabine's magnetic odyssey, in post-Napoleonic peacetime, the British Admiralty had sought a new attention-getting mission for its unemployed officers. Fatefully for many, they settled on Arctic exploration. Sabine, a gifted opportunist, immediately attached himself to Edward Parry's 1819 search for the Northwest Passage, the first to overwinter in the Arctic, as its unofficial magnetic ambassador. For two years, he took readings in the fjords of Greenland, on top of icebergs in Davis Strait, and along majestic Lancaster Sound. He estimated the location of the North Magnetic Pole, and witnessed the aurora borealis whip his instruments into a frenzy.

On his return, the Royal Society promptly shipped him on a circum-Atlantic voyage to fill in the magnetic map of the Northern Hemisphere. For the next three years, the irrepressible Edward Sabine might be found peering at his dip circle all across the Atlantic trade zone, in sequestered locations free of iron. He scribbled readings in a palm grove in Porto Praya, on a beach in Tenerife, beneath fortress walls in Sierra Leone, and on the grounds of the lunatic asylum in Manhattan in a snowstorm. The data were prolific but inconclusive. The physics committee at the Royal Society noted the weird simultaneity of magnetic anomalies across distant locations, but also that variations in magnetic inclination, dip, and intensity "apparently observe no law."

Five years into his magnetic crusade, Sabine was only just warming to the task. An entire hemisphere remained to be data mined. For the Victorian polar aspirants Ross, D'Urville, and Wilkes, the vast blank of the southern high latitudes offered a last chance at glory before the world was filled in. For Sabine, by contrast, undiscovered coastlines were less compelling than their electric envelope—the ineffable airborne map of the magnetic field. If charting the Antarctic continent promised the rival explorers instant promotion to the ranks of Cook and Magellan, then Sabine, were he to discover the laws of magnetism from data gathered at the South Pole, might yet be hailed as a second Newton.

But stiff political headwinds faced a magnetic expedition to the South Pole in the 1820s. A naval attempt on Antarctica would require massive funds, for a mission far outside the British networks of trade and with little prospective return on investment. Years passed. Sabine lobbied the Royal Society to no avail. So he switched allegiance to the progressive British Association for the Advancement for Science, whose high-profile annual meetings offered a platform for evangelizing the magnetic cause. An Antarctic voyage was needed, Sabine argued, to ensure Britain's leadership in international science, under threat from the celebrated magneticians of France and Germany. As for utility, the practical benefits of discovering magnetism's universal laws were only uncertain in the sense they surpassed imagination.

Sabine's petitions were ignored. So, with patriotism a losing argument, he turned to the more subtle power of celebrity, whose influence crossed borders. Ironically then, it was the consummate internationalist, Humboldt, who came to the rescue of British magneticians in the spring of 1836 with a letter written, at Sabine's behest, to the president of the Royal Society, the Duke of Sussex. The famous Humboldt letter, written in French, mixes impassioned argument for magnetism's importance to "the progress of human knowledge" with liberal doses of flattery and aristocratic name-dropping. Magnetic data gathering required physical research stations in the four corners of the globe—in other words, Britain's colonies. The empires of land and the new science being coextensive, the young queen lacked only a suite of magnetic observatories across her far-flung dominions to complete her world-imperial inventory. This included the greatest magnetic prize of all: an observatory at the South Pole (British by right).

Humboldt's authority as the founder of modern magnetic science, and friend of kings, turned the tide in Sabine's favor. The Magnetic Crusade was already, in the words of Royal Society power broker William Whewell, "by far the greatest scientific undertaking the world has ever seen." Now, with the timely intercession of its Prussian figurehead, the sputtering endeavor was revived. Edward Sabine's decade-old proposal at last had the attention of Whitehall. The prime minister signed off; the chancellor released the funds; and the modern major-general had his South Pole expedition with a readymade polar hero as his proxy. Crossing the Antarctic Circle into the electric unknown, James Clark Ross and his magnetic crusaders would soon be at Jerusalem's gates.

Once in Hobart, Ross's orders were to erect a permanent observatory, fully staffed, for hourly readings. A Tasmanian observatory, the reasoning went, would help fill vast data gaps existing in the Southern Hemisphere and establish, at long last, a universal law for the enigmatic operations of terrestrial magnetism. More practically for the British government, since the annual number of ships lost to bogus compass readings stood scandalously high, a magnetic breakthrough in navigation technology would protect trade and guarantee Britain's dominion over the oceans.

For the site of the observatory, Ross and his fellow Arctic veteran, Governor Franklin, set out from breakfast one morning to survey a quarried bed of sandstone on the grounds of Government House, notably free of iron-bearing rocks. By that same afternoon, two hundred convicts had arrived to dig foundations, erect pillars, and drag timber to the site. In place of nails, they fitted specially crafted wooden pegs. The interior of the observatory was (needlessly) lined to reduce temperature fluctuations from outside, while a wooden partition shielded the magnetometer from the deranging body heat of the magnetician, who would read off the tiny numbers on the instrument while peering through the lens of a telescope mounted on a step.

Perched like a bulky wooden lighthouse overlooking the picturesque Derwent River, the Hobart structure joined a legion of magnetic observatories already operating globally from Saint Petersburg to Peking. This was the first iteration of a magnetic consortium that still exists today, when more than two hundred observatories worldwide are dedicated to tracking Earth's magnetic field. To honor the international spirit of the Magnetic Crusade, Ross suggested the new observatory be named after Carl Friedrich Gauss, a leading German theorist. But Lady Jane Franklin wanted it named for her hero, and so it became "Rossbank." Hourly observations began, with round-the-clock shifts. Captains Ross and Crozier slung hammocks from the rafters, for naps between magnetic readings. For third-placed James Ross, a man under pressure, Sabine's magnetic toys were a welcome distraction from brooding over Wilkes and D'Urville, and sleep a welcome release from the deadweight fear of shaming the nation. His waking hours, meanwhile, were consumed in plotting his risky southward course, his inspiration for which lay in a lucky encounter back in England a year prior.

Back in September 1839, just days before the *Erebus* and *Terror* had been due to sail from Chatham docks, Captain Ross's steward had come below to pass along the request of a Mr. Charles Enderby to come aboard and speak with him. The visitor had inherited the firm of Samuel Enderby and Sons (featured in *Moby-Dick*) from his father ten years before, but the son's dreams were not profit driven. Rather, he wished to be a great patron of exploration, and to be accepted by the lords and intellectuals of the Royal Geographical Society as one of their own.

Enderby the younger had commissioned James Weddell on his record-breaking journey, in addition to two other costly expeditions to the polar south, of which the first—captained by John Biscoe—had charted fragments of land to the north and east of Weddell's path. The third and most recent mission had returned only days before, and Charles Enderby was grateful he had the opportunity to pass along the logs and charts from that voyage to the famous Captain Ross. John Balleny, the man Enderby sent, taking a direct poleward course from New Zealand, had discovered icebound islands and a volcano, the southernmost land ever seen at that longitude. Moreover, Balleny thought—*thought*—he saw the "appearance of land," perhaps a continent, even farther to the south, before disaster had struck.

John Balleny had been well over sixty years of age, and retired from the sea, when poverty induced him to answer Charles Enderby's call for a mission of southern discovery. His sole consort was the tiny cutter *Sabrina*, commanded by a man named Thomas Freeman. Noting the absence of harpoons and whale lines on deck, the crew of the *Eliza Scott*—recruited by who knew what means—concluded that profits from this voyage were bound to be slim, and behaved accordingly. Balleny's first mate abused him for dismissing all women from below deck, drank heavily, and sulked the entire journey. On the first Sunday at sea, when the captain called divine service, the entire starboard watch refused to attend. Soon Balleny had only a half dozen men he could rely on.

On the first day of February 1839, the *Eliza Scott* and *Sabrina* reached the ice pack at 69° south, the highest latitude any ship had ever traveled along that meridian. For several days, they tacked along the rim of the pack in a dense fog. On February 9, the sun broke over the deck and they could make out dark shapes on the horizon, which evolved into ice-covered islands, the bare rock showing from where icebergs had calved off. Heaving to as close as they dared, Freeman joined Balleny in the *Sabrina*'s boat, to make landing. But a shift in the waves immersed the beach so that Freeman was left waist deep in the freezing water, clutching a few samples of rock. Behind him, in the distance, a volcanic island sent tendrils of white smoke into the dull sky.

In the first days of March had come the tantalizing glimpse of a coast, visible beyond the ice. But the little schooner and cutter could not hope to penetrate the pack, and their "discovery" remained out of reach. An enormous iceberg passed with a block of dark stone embedded in it: further proof of some great land in the offing. That same day, Freeman crossed over to the *Eliza Scott* in *Sabrina*'s boat, bringing with him his ship's boy, named Smith, whose insolence had driven him to the brink of homicide. Balleny exchanged Smith for a milder boy from the *Eliza Scott*, named Juggins, who shipped off with Freeman back to *Sabrina*. Balleny ordered Smith aft to the tiller, but Smith cursed at him, let go of the tiller, and threw a rope in the captain's face. Balleny, for all his sixty years, was a match for the delinquent. He grabbed him by the throat, drove him forward, and beat him bloody. When, on March 24, the *Sabrina* sank with all hands in a gale—last seen as her blue distress lights flashed through the surf—Balleny's first thought was for poor Juggins, whom he had switched for the worthless Smith. The coast they spotted in the distance through the spray now bears the name of the lost cutter.

James Ross listened intently to the story of Balleny's voyage. This polar maniac Enderby had sent two frail yachts—a schooner and a cutter—into the most dangerous unexplored waters on Earth. He guessed that the Enderby fortunes must have fallen to a low ebb, since it required seven merchant backers to finance even this meager flotilla. It was well known in seafaring circles that the Biscoe expedition had cost tens of thousands of pounds, with very little return in the form of whale oil and sealskin. And the *Eliza Scott* had limped home alone with only a cluster of vaguely charted islands to her credit.

But Ross instantly saw the value of Enderby's investment. The old captain's glimpse of land beyond the pack, at a longitude south of New Zealand, marked an alternate route to the pole. *He* had the ships, the mighty *Erebus* and *Terror,* to ram through that pack, navigate to the hidden coast, and raise the flag. (Balleny's unlikely voyage paved the way not only for Ross, and later Scott and Shackleton—today, the Balleny corridor marks the main shipping route to the scientific research bases of the Ross Sea.) As for Charles Enderby, his polar obsession ultimately brought the family's whaling empire to its knees. He lived out his days billeted with his daughter and her family in cramped rooms off Green Park in London, where it was the first rule among the Enderbys never to lend money to Uncle Charles.

Entering the harbor of a lush, sub-Antarctic island named for Charles Enderby in late November 1840, the *Erebus* and *Terror* encountered a violent squall, howling down from the hills to the west. It required five hours, with all hands, to bring the ships safely to anchor. On the beach, they were met with two painted boards erected in the sand. On one, a painted note from the Americans: the brig *Porpoise*, out of Sydney, had landed at Enderby Island eight months prior, on her return from an exploring cruise along the Antarctic Circle. A message in a bottle gave further details, though the writing was blurred by water stains. The *Porpoise* had cruised along the ice pack, presumably on a more northerly track than its flagship *Vincennes*. Neither the sign nor the message in the bottle made mention of Wilkes's vaunted "Antarctic Continent."

The other board, however, was more discomforting: a communication in French from Dumont D'Urville. The *Astrolabe* and *Zélée*, sailing from Hobart, had landed at this spot on March 11 (barely missing the Americans), before departing for New Zealand. On January 19, it announced, the French ships had discovered "Terre Adélie" and determined the location of the South Magnetic Pole.

Ross's heart sank. More unsettling even than D'Urville's discovery of land was his boast about the magnetic pole. To concede both a continent and the pole to the French would mean sailing home in disgrace. He would rot on shore on half pay—and never get married. The English captain's hopes rested only with D'Urville's vague wording. He had "determined" the pole, not reached it.

In a sheltered cove on the western side of the harbor, they landed the portable wooden observatories. Sabine's ambitious magnetic calendar designated "term days," when observatories around the world agreed to take magnetic readings every two and a half minutes during a twenty-four-hour period. If regular hourly observations produced mostly noise, term-day readings of variation at close intervals might reveal the longed for pattern of a universal magnetic law. Term day was approaching, so Ross ordered all hands to clear away trees and dig foundations for emplacement of the observatories. The deeper they dug, the softer the peat bog became, so they threw down blocks of stone and casks filled with sand as base for securing the magnetometers.

Sabine had speculated that magnetic activity in the southern high latitudes might be totally original, and so provide a breakthrough clue. But the Enderby Island data offered no tantalizing outliers. Flickering in instantaneous response to the magnetic impulses from Earth's liquid core, the instruments behaved in every respect as they had during Ross and Sabine's years of lonely observation in the Arctic.

In addition to an uninterrupted week of readings, Sabine had directed Ross to make absolute determinations of magnetic inclination, dip, and intensity at Enderby Island. Here, Ross failed. During their cruise from Hobart to Enderby, Fox's dip circle showed a steady increase in dip in exact proportion to the distance they sailed toward the pole. But once on the island, the instruments went haywire. Readings thirty yards apart differed wildly. Approaching the cliffside rocks, the compass needle began literally to spin. When he tested the magnetism of the rocks themselves, Ross discovered their polarity varied from north to south depending on the random position in which he found them. McCormick, his geologist, brought him a bag of ferruginous samples from all across the island. Enderby, Ross concluded, was "one great magnet," and a magnetic island rendered his instruments useless.

Quitting Enderby Island, the *Erebus* and *Terror* passed the Antarctic Convergence, where tropical waters from the north are absorbed in the great conveyor currents of the cold Southern Ocean. On New Year's Eve, they crossed the Antarctic Circle, whereupon they abrupted immediately at the ice pack, lilting and crackling through the heavy snow. Ross was at first dismayed to meet with the pack so far north but took heart at its thin, modular appearance. When the weather cleared, the officers made a full appraisal. The glinting, ice-crusted ocean extended as far south as could be seen from the nest, without the least hint of land or other termination. Somewhere in that great floating ice tundra, Ross thought, lay his prize: the South Magnetic Pole. But would his ships survive the pack to get the crews there, and then get them back out?

Ross brooded at the brink, eyeing his opportunity. The dip circle read 81°40′, less than nine degrees shy of the pole. Another

Fig. 11.1. The enormous risks facing Victorian sailing ships in stormy, ice-strewn seas is captured in this image of HMS *Erebus* and *Terror* in the Ross Sea.—James Ross, *A Voyage of Discovery and Research in the Southern and Antarctic Regions* (1847). Rare Book and Manuscript Library, University of Illinois at Urbana-Champaign.

snowstorm brought heavy seas, launching loose ice across the decks in sheets of freezing foam. Icebergs appeared through the gloom, rakingly close. On January 5, Ross skirted the pack again, probing for weakness like a bear-baiting dog. They came to an inviting inlet of open water, and Ross felt the eyes of sixty men turn to him as one. A stiff wind was blowing in from the north, directly onto the ice. If the ships took the pack now and fell into trouble, there could be no return to the safety of the open ocean. A responsible captain would wait for the storm clouds to pass, at the very least, before ordering his wooden ships into a treacherous sea of ice as hard as rocks, with no discernable end or route of escape.

Ross didn't hesitate.

Interlude: The Magnetic Crusaders

Before Douglas Mawson, following in the wake of Charles Wilkes and Dumont D'Urville, led his own expedition to the gale-blown coast of East Antarctica in 1912, he sailed under Ernest Shackleton aboard the *Nimrod* to West Antarctica and the Ross Sea. From the little explorers' hut at Cape Royds, beneath Mount Erebus, Shackleton launched his first unsuccessful assault on the geographic South Pole, while Mawson set out with two others in the opposite direction (north). Their orders were to make a geological reconnaissance of the exposed mountain shelves of Victoria Land, and to honor the memory of compatriot James Clark Ross by setting foot at the elusive South Magnetic Pole.

Back in 1831, Ross had stood on the Arctic tundra and watched the fragile needle of his dip circle drop until it pointed directly past his feet, through the center of the Earth, to its imaginary opposite point in the Antarctic ice. The joy Ross felt in that moment made a permanent impression. Thereafter, magnetic south loomed for him as the ultimate unclaimed prize of exploration. He would be the first to stand at both magnetic cynosures of Earth.

For Shackleton and his Edwardian rivals, by contrast, magnetism held little attraction. The geographic pole offered a definitive compass point on the ice—the location for a triumphant flag and photo opportunity—while the magnetic pole seemed a more academic, nineteenth-century obsession. The polar site of vertical magnetic intensity on Earth shifts from year to year and even hour by hour, as obscure as a scrubbed out equation on a chalkboard.

Magnetism's high stakes, let alone its mathematics, were a hard sell to the early-twentieth-century public and to expedition patrons.

On the *Nimrod* expedition, therefore, Shackleton delegated this B-list adventure to men who could be spared from his epic South Pole campaign. A spindly middle-aged professor named Edgeworth David was to plant the Union Jack as best he could at the magnetic pole beyond the Transantarctic Mountains, with two sturdy younger companions, Mawson and Alistair Mackay, as travel insurance. No one blamed Shackleton for this bare-bones organization. But at the lowest moments of their four-month odyssey, the geologist trio did have reason to curse the memory of James Ross who, recalling his long experience sledding the low-lying ice plains of the Canadian Arctic, had considered an overland trek to the South Magnetic Pole a simple task.

In the waning days of the 1908 Antarctic summer, Mawson, Mackay, and the professor had viewed the course of their prospective journey from the summit of Mount Erebus. They were the first to make the ascent. Through gaps in the steam clouds, the spectacular landmarks of Ross's famous 1841 voyage lay spread before them—a scenography for Antarctic exploration's Heroic Age.

Snow-clad Mount Erebus, bordered by its companion volcano Mount Terror, stood like smoking sentinels at the western limit of the Great Ice Barrier (later called the Ross Ice Shelf). Rocky moraines at a thousand feet showed that the ice barrier had once engulfed the volcano slopes and was now in retreat. At the eastern foot of Mount Terror, at Cape Crozier, swarmed the largest colony of emperor penguins in the world. During Robert Scott's winter residence at Cape Evans a few years later, Cape Crozier was the object of the most harebrained polar quest of all time, when three men traversed the mountain slopes of Erebus and Terror, in the howling dark of winter, to retrieve a single emperor penguin's egg (a trek immortalized in Cherry Apsley-Garrard's narrative, *The Worst Journey in the World*).

Erebus's crater, a juncture of fire and ice, hissed like a giant steam bath. At intervals, the half-mile-wide cauldron boomed, and Mawson and the others were instantly enveloped in a sulfurous cloud. From the summit, looking west at sunrise across ice-strewn McMurdo Sound, they witnessed an extraordinary sight. A conical shadow image of mighty Erebus raised itself, forty miles wide, on the rocky slopes of distant Victoria Land, the northern reach of the Transantarctic Mountains. Their route to the South Magnetic Pole lay along that icebound coast, from the Ferrar Glacier north toward Coulman Island (named by James Ross as a peace offering to his future father-in-law). Mawson's dip circle would direct them west up the mountainside to a high inland plateau. Somewhere across that remote ice expanse, the magnetic pole, their destination, was drifting, as inconstant as the Erebus vapor cloud swirling around them.

They set out in early October with a motorcar pulling the sled heaped with provisions. They brought ponies, and a gramophone to play sentimental favorites in the evenings. But these were quickly abandoned along the way, and soon it was three explorers manhauling a stacked sled across a brutal icy plain bordering the Ross Sea. Most days, they couldn't tell if the ice they trod across rested on land or water.

At the mouth of the Ferrar Glacier, named for Scott's geologist on the *Discovery*, they paused to contemplate the magnificent geometry of the Transantarctic glacial valleys. On either side, chocolate-colored crests of Ferrar dolerite ran in perfect parallel, merging with the ice at the elevated horizon in reddish purple tints. The spectacular Ferrar sills, comprising seven thousand cubic kilometers of magmatic rock, ran in thick terraces north and south as far the eye could see—the closest equivalent on Earth to the landscape of Mars. Their creation, one hundred eighty million years ago, marked one of the most intense volcanic events in Earth history. On the mountainsides, they observed distinct furrows in the exposed rock—marks of a mega-glacier's progress at its maximum bulk, presumably recent (in geological time). A little farther north, they unpacked their flag and, despite being more or less helpless dots in a hostile landscape, claimed possession of the Transantarctic Mountains for King Edward and the empire.

The beginning of November found them still struggling along the coast, dragging the sled up and down endless ridges of bare blue ice. When they came to a chasm, the sled had to be unloaded, transported by rope across the divide, then repacked. Mawson took out his dip circle, which showed them hundreds of miles southeast of the South Magnetic Pole. It was now clear they could not reach the pole and return on full rations. So they went hungry and grew weaker, husbanding provisions for the backbreaking ascent to the plateau. They began to hallucinate about food and named a prominent cape after the expedition cook. Meanwhile, Mackay was developing a violent resentment toward Edgeworth David, kicking at his heels from his rear position in the harness. The fifty-year-old professor could not pull his weight and seemed to be slipping in and out of dementia.

Two interminable months later, up on the plateau, even reality took on the character of a hallucination. The boundless sea of crystallized ice, called névé, glittered like diamonds. One morning, they observed three distinct suns in the sky. Reflection on airborne crystals produced a glowing halo at the horizon, where the rims of two mock suns shimmered alongside the true orb. Meanwhile, snow blindness made readings of the magnetic dip circle increasingly difficult. In intervals of calm, Mawson ascertained that by contrast to the 1841 magnetic readings at this latitude, when Ross observed the magnetic pole traveling easterly toward their present location, it was now moving northwest, away from them.

The skin of Mawson's lips had peeled fully away, leaving raw flesh. In the mornings, he woke to find his mouth sealed shut with dried blood. Mackay was sullen, and the professor intermittently raving, but fear of failure trumped all trauma, so they kept on. On January 15, the dip circle read 89°45′, a hairsbreadth from vertical. At that moment, they were mere miles from their goal. Given daily magnetic wandering, Mawson reasoned, they might as well stay where they were and wait for the pole to pass beneath them. But mathematical logic did not satisfy their explorers' sense of occasion, so they left their tent and walked five miles in the approximate direction of the pole. At latitude 72°25′, thousands of miles from geographic south, Mawson signaled their "arrival" at the magnetic pole, calculated as a mean position based on pages of scribbled readings. The photograph commemorating their achievement shows a sturdy flag on a wooden pole, and a trio of indistinguishable men with unsmiling, blasted faces. Their heads are bared out of respect for the moment of conquest. Behind them, a freshly fallen snow merges into the white sky. The discoverers of the South Magnetic Pole look like human smudge marks on a blank sheet of paper.

In 1836, Alexander von Humboldt had described the research scope of terrestrial magnetism to the Duke of Sussex, president of the Royal Society, as "the work of centuries." His estimation was not greatly exaggerated. With Ross's return from the Southern Ocean in 1843, Edward Sabine had logbooks containing hundreds of thousands of magnetic readings from all over the world. He took command of a brigade of army clerks in an office building in Woolwich to process the data through several million separate mental calculations. This required a decade. But even after all this effort and expense, the reams of figures revealed no patterns, no order, and certainly no universal magnetic law. In times of doubt, Edward Sabine wondered if his thirty-year Magnetic Crusade had been one giant folly.

Just in time to save his reputation and career came a lucky break. Sabine's wife, Elizabeth, a gifted linguist, was translating Humboldt's epic volume *Cosmos* for a London publisher. Looking over proofs at home, Sabine learned of the work of Humboldt's compatriot Heinrich Schwabe, who had dedicated his life (with a fanaticism Sabine could relate to) to recording sunspot activity. To his astonishment, spikes in Schwabe's sunspots matched the magnetic anomalies, or "storms," in his own terrestrial data. Earth's interior forces driving the magnetic field remained mysterious, but here was proof of magnetism's essential role in solar-terrestrial dynamics and the electrical connection of planetary bodies.

Sabine and Ross did not live to see magnetism's full potential realized, but their geophysical instincts were correct, and their herculean efforts proved invaluable to the history of Earth science. Magnetic tools and data, developed from the mid-twentieth century on, proved decisive in tackling the great unanswered questions of Earth's, and Antarctica's, history.

In January 1954, a group of British magneticians gathered for a meeting in Birmingham. The American magnetic research establishment—based at the Department of Terrestrial Magnetism in Washington, DC—had grown stagnant, but postwar tyros in England found energy as a loose, interdisciplinary collective. These next-generation magnetic crusaders began with the same grand problem that had driven Sabine, Ross, and Mawson: namely, the baffling complexity of Earth's magnetic field, produced by the roiling dynamo of Earth's liquid core, and interfacing with solar winds. Their renewed investigations soon led them from the unfashionable backwater of terrestrial magnetism into the glare of twin major controversies driving twentieth-century science: continental drift and climate change.

At the Birmingham meeting, a parade of speakers expatiated on the emerging field of paleomagnetism: how a wide variety of rocks, in their rapid cooling, carried a permanent record of the direction of Earth's magnetic field at the time of their creation, and thereby opened a window onto magnetic variation over geological time scales. A hallmark of early paleomagnetic research was the head-spinning data and their planetary implications. First and foremost, Earth's poles had flipped 180° numerous times. In the spirit of Edward Sabine, the speakers in Birmingham brought out their paleomagnetic travel diaries. From the volcanic beds of Iceland to the isle of Mull in Scotland to the Columbia River in the northwest United States, rock samples all showed intermittent *reversals* of Earth's magnetic field. In addition, within each polar regime, magnetic direction varied, confirming the magnetic poles were not fixed but prone to wander.

One speaker, James Clegg, even dared to mention continental drift, then an almost heretical doctrine. If polar wandering was insufficient to explain magnetic obliquity in rocks, it must be the rocks themselves that moved. In the animated discussion, it was agreed that only comparison of paleomagnetic data from different continents could solve that issue. If landmasses changed positions relative to each other, as Alfred Wegener had controversially proposed back in the 1920s, the data would show it.

One attendee at Birmingham, graduate student Ted Irving, was determined to publish first. Despite needing to rush to finish his dissertation while also packing his bags for a job in Australia, Irving still managed to gather vital data from correspondents in India and Tasmania. As he had dared to guess, the "position" of the pole ranged wildly between continents. The Asian results, combined with the data of his colleagues in Birmingham, showed that Europe and North America had drifted apart since the Jurassic, and that India, most dramatically, had once been stationed in the Southern Hemisphere before drifting north to its present position, spinning slowly as it went.

As a young scholar in a field bitterly divided by Wegener's "mobilist" theories, Irving understood the professional danger in his claim. He might have proved continental drift, but he could not bring himself to name it. His breakthrough 1956 article thus carries the coy title, "Palaeomagnetic and Palaeoclimatological Aspects of Polar Wandering," and does not mention drift. Reviewers at the Geological Society of Australia penetrated the ruse instantly and rejected the paper, forcing Irving to publish his scoop, almost surreptitiously, in an Italian journal.

When, five years later, Fred Vine and Drummond Matthews at Cambridge identified "stripes" of alternating magnetic polarity in the seafloor, proof of its dynamic outward spread, continental drift completed its evolution into plate tectonics and ultimately won the day. The American geological establishment—stubborn champions of continental fixity—grudgingly capitulated. But victory was never inevitable. Prior to the British magnetic revival of the 1950s, Wegener's continental drift had been mostly ignored. Via paleomagnetism, plate tectonics emerged within a short decade as the foundation of modern Earth science. A century and more after the first magnetic crusade, Humboldt, Sabine, and James Ross enjoyed a belated but monumental vindication.

In addition to paleomagnetic proof of continental drift, Ted Irving's 1956 paper was the first to connect rock magnetism to paleoclimate. The magnetic birthmark of rocks included their original "dip," or inclination, which pinpointed their latitude at the time of formation. This, in turn, could be shown to correspond to the climatic signatures of the sediments within which the rocks were embedded. And because rocks were datable within a larger timeline of polar reversals, a rich parallel climate history emerged from the magneto-stratigraphic column, interwoven with the dance of continents, the flux of oceans, and the rise and retreat of the great ice sheets.

Most importantly, the establishment of the Geomagnetic Polarity Time Scale (GPTS) provided a time clock for marine oxygen isotope analysis, by which past polar ice sheets could be tracked and measured. It was through isotope analysis that Jim Zachos and the scientists of Leg 128 had identified the Eocene-Oligocene Transition, thirty-four million years ago, as the global cooling episode in which a continental ice sheet first engulfed the Antarctic landmass. But doubts lingered. The fossilized shells of ancient foraminifera buried in ocean beds were proxy evidence for glaciation, but could *direct* physical evidence be found for Antarctica's first ice?

Enter magnetism once again, in its latest guise as the handmaiden of paleoclimatology. In the early 1970s, it was discovered that magnetic minerals, found worldwide from lake beds to desert outcrops, offered a detailed archive of global environmental change over millions of years. The most ubiquitous magnetic mineral, called magnetite (the legendary lodestone), originates in the same basaltic lava flows that, fulminating beneath the ocean, allow the dating of tectonic plates. When the iron-rich crystals of basaltic lava cool, magnetic crystals—black, shiny, and dense—are sealed in rock beds. Over geological time, wind and rain weather the rocks, releasing the mineral grains on a second odyssey via the air or river stream to a new, vagrant residence as sediment. Dug up and scrutinized in the lab, the magnetite grains reveal a detailed history of their adventures and of planetary climate change. Magnetite intensity and concentration within the sedimentary layer are hieroglyphs of glaciation. For example, magnetite deposited in low concentrations might signify the cold, dry climate of an ice age, while the humidity of an interglacial period delivers magnetite to the ocean basin at an exponentially higher rate.

For Antarctic climate historians, and the dicey question of first ice, magnetite was the mother lode. In late 1986, a New Zealand research vessel retraced the course of the *Erebus* and *Terror* southward through the ice pack, past windy Cape Adare, and into the Ross Sea. With the Transantarctic mountain range close off the starboard bow, the ship mimicked the route of Shackleton's *Nimrod* in 1909, desperate for a glimpse onshore of its three missing seekers of the South Magnetic Pole (they found them alive, by pure chance). At the outlet of Ferrar Glacier, 77° south, where Douglas Mawson and his companions had begun their historic quest, the latter-day magnetic crusaders established a drilling platform on the ice pack, thick enough to support more than fifty tonnes of industrial machinery for a month. They installed hydrocarbon sensors in the drill core to avoid stumbling on a flammable gas deposit and immolating the whole operation. Drill core samples were flown daily by helicopter to the nearby American station at McMurdo Bay, where the paleomagnetists enjoyed their own dedicated lab. Moist sections of the core were brought to them like religious offerings, on trays.

In the lab at McMurdo, the technicians determined that the magnetite grains extracted just offshore had originated in the Jurassic period, during a convulsive volcanic period that precipitated the breakup of Gondwana and the creation of the Transantarctic Mountains. Over millions of years, wind and water shaped the valleys and fluvial courses by which the weathered magnetic mineral, mostly Ferrar dolerite, made its way from the mountains to the sea. Then came the ice, sealing off the rich volcanic soils and wooded alpine terrain like a sarcophagus.

Later, an international team of paleomagnetists, led by Leonardo Sagnotti in Rome, determined that magnetite levels in the first core, called CIROS-1, fluctuated dramatically according to their situation in the sedimentary column. At a critical point, four hundred thirty meters below sea level, magnetite concentration changed abruptly. Beneath four hundred thirty meters, Sagnotti concluded, the Ross Sea climate had been warm and humid, weathering abundant dolerite and mixing magnetite with soils washed from the slopes of the Transantarctic Mountains. Above four hundred thirty meters, by contrast, magnetite levels dropped precipitously, and the grains themselves appeared finer and finer upward through the column. This was the work of an arid glacial climate, resembling Antarctica today, that limited erosion and ground what magnetite remained into ever smaller particles.

The breakthrough achieved by the CIROS-1 core was to confirm the date of Antarctica's initial glaciation at the Eocene-Oligocene Transition, a timeline first proposed by Jim Zachos and the scientists of Leg 120 of the Ocean Drilling Program in 1998. Four hundred thirty meters below sea level corresponded precisely with the EOT; it was the farthest south this critical Earth history boundary had been identified. Physical evidence of glaciation on the Victoria Land coast, in the heart of the Antarctica, certified what microscopic fossils buried off Kerguelen Island, thousands of miles distant, had suggested by proxy: Antarctica's ice sheet originated thirty-four million years ago, in the midst of a grand climatic upheaval that revolutionized the biosphere. Magnetite grains buried beneath the Ross Sea signaled the genesis of our Glacial Earth. The Magnetic Crusade, launched in Victorian times with Edward Sabine and James Ross, had been pursued across generations. Setting out in search of one holy grail—magnetism's nonexistent universal law—they had found another even more marvelous.

⸸ 12 ⸸
Ross in Wonderland

When James Ross decided to charge the pack in early January 1841, the *Erebus* and *Terror* were a mile distant, with a strong breeze from the north. He reduced sail to deaden the ships' way, and eased them against the ice edge. Once within the rim, Ross hoisted sail to force the issue. Soon, they could no longer see water at any horizon, and were entirely hemmed in by the pack. Having crossed the Antarctic Circle and now into the pack ice, Ross ordered the cold-weather gear to be distributed: box jackets, boots, thick stockings, and Welsh wigs (a kind of woolen skullcap). The ice was flat and snowy white, with scattered hummocks. Seals regarded the crews listlessly from the floes, while the men in the rigging remained on alert for whenever a stretch of water opened briefly to the south. Otherwise, they held on for dear life as the hull bashed again and again against ice blocks the size of pontoons. The sound of the hull splintering and cracking sickened them.

On January 6, the officers celebrated twelfth night in the captain's cabin in customary style: with cake, riddles, and impersonations of royalty. Then, like a belated yuletide gift, the pack gave way to open water, with more dark water sky, suggestive of ice-free conditions, ahead. With the dip circle at 85°, *Erebus* headed directly for the South Magnetic Pole, which they hoped to sail across within days. But James Ross was not destined for two prizes. His daring route south through the pack

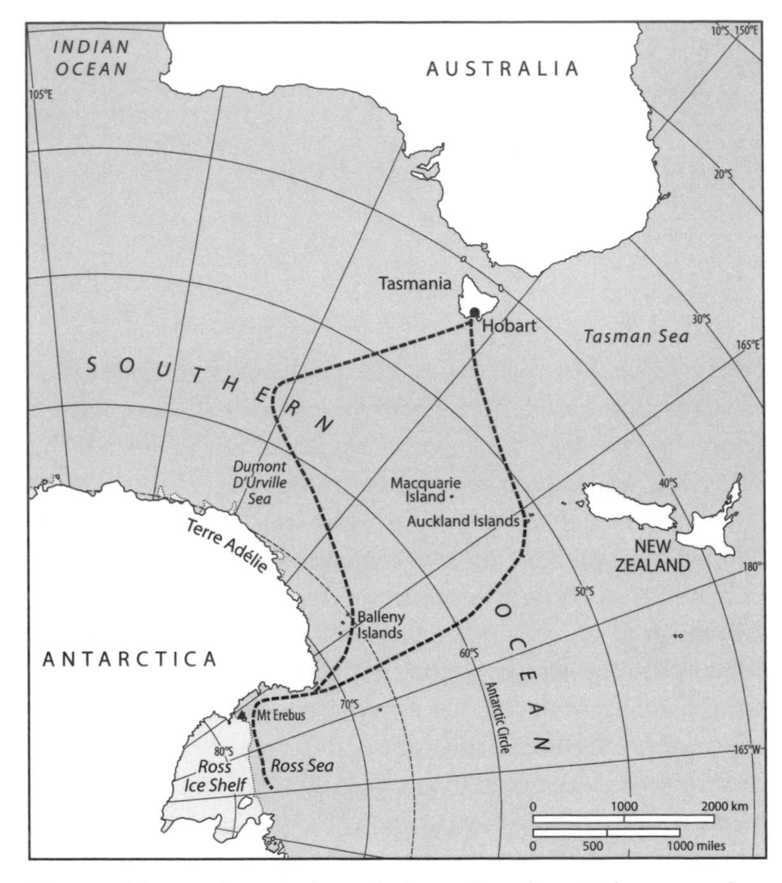

Fig. 12.1. The novel course chosen by James Ross from Hobart opened up the southernmost polar sea for the British to explore in the summer of 1841. Ross's epic voyage blazed the trail for Antarctic exploration's "Heroic Age," launched at the end of the nineteenth century.

brought him Terra Australis, an undiscovered sea, and a multitude of wonders, but not the pole.

A northerly gale had blown them to the brink of a castellated glacial kingdom none of them, even the Arctic veterans, had thought to imagine. The first sight of land—a range of snow-capped peaks clear and cold—dissolved into cloud. Then, at two in the morning, a new formation appeared through the haze. The off-duty officers, roused from their beds, crowded the port bow, squinting through their spyglasses. For an hour, the conical shapes became so intermingled with the misty drift and play of light, they believed they had fooled themselves again.

Then the cloud bank lifted to reveal proper mountains dead ahead, with floating white islands at their base. Ice and snow blanketed the range in a vast, crystal mass, except for the rocky peaks, or where sheer black cliffs abrupted directly at the water's edge. The weather cleared to create a magnificent graduated triptych: gloomy cliffs, white peaks, and bright blue sky. Like the accidental audience to a vision, the men of the *Erebus* and *Terror* lined the bows in hushed silence, while Captains Ross and Crozier skirted the deadly shore with instinctive care.

The westward-cleaving mountain slopes, pathway to the pole, stretched to infinity. Nestled between them, miles-wide glaciers, descending who-knew-how-far from the unseen interior of the continent, crowded the valleys and overflowed the coast, forming ice shelves miles long. After breakfast, the ships came to the tallest summit in the range, rising steeply from the shore until it pierced the clouds. Ross named it Mount Sabine, because it must overlook the South Magnetic Pole. The great mountain cluster, stretching toward the south, he christened Admiralty Ranges. For the British sailors, even the officers familiar with alpine scenes, the scale of the range stunned them. So, too, did the sheer quantity of ice, far beyond anything on the Arctic tundra.

Leaning on the rail, sketch paper in hand, Joseph Hooker called to mind descriptions he had read of the Himalayas and the Andes, only subsumed entirely by ice. Nothing in his native Highlands compared to the sight of these giant, crystalline precipices, crowned by golden clouds against a never-setting sun. Above these, he attempted to capture the unnatural blue sky, free of dust and vapor, which framed the scene so that every detail seemed impossibly close and real. Higher again still, he drew in the bank of dark clouds, like charcoal, suspended along a crimson rim of light. The landscape was almost too glorious. And the human eye was an inadequate lens. Seeing it was like spoiling it.

Putting the sketch away, Hooker called for the dredge to be put over the side. Back in England, naturalists at the Royal Society had assured him no marine life could survive in the polar cold, especially at depth. But the contents of the net overthrew all such assumptions made in temperate latitudes, disgorging heaps of mollusks, shrimps, sea butterflies, a red crustacean, and brightly colored corals onto the deck. The beneficiaries of this sea-world cornucopia, the whales and petrels, swarmed above and around the *Erebus*. While the mountains ahead loomed still and austere as a painting, their marine surroundings teemed with life—noisy, active, and rapacious.

By noon, the ships had passed Captain Cook's farthest point south, 71°15′, still with open sea to the horizon. The magnetic pole, Ross reckoned, lay at 76° south, five hundred miles to the south and west, on the other side of the mountains. Faced with the choice of sailing north or south to find a circuitous route to the pole, Ross chose south. If necessary, this new southern land must vindicate him. He and his men would take a beach somewhere, plant flags, establish a base, explore, and stay the winter. That was the plan. But even this consolation prize, the southernmost land ever seen, almost within touching distance, would keep him frustratingly at bay. Before the end of their cruise, Ross had ordered the designated discovery flags to be run up the mast for an airing, to prevent mold.

On January 12, Ross first signaled to *Terror* his intention to land. The men lowered the boats and, on reaching a cluster of islands a mile from shore, found themselves drawn into a fierce current. The agitated swell bombarded the boats with ice, then almost swamped them. The cold wind off the glacier took their breath away. To attempt landing on the "shore," actually a ten-foot-high cliff of ice, hard as rock, was out of the question, so they stopped pulling against the current, and settled for a small island full of penguins to make their conquest of Antarctica. Its beach, too, was walled by ice, but Ross, choosing his moment, threw himself off the boat and scrambled over—the first to set foot. Crozier followed, and then the rest. The island reeked with the rotting corpses of penguins and the thigh-high excrement of their living brethren. Climbing to the island's miniature peak, the explorers sank into it like a bog. They raised the flag, while Ross gave a short speech, claiming everything in sight for Queen Victoria. Then they gave three cheers, as the penguins screamed and hacked at them with their beaks. The Royal Navy's occupation of Possession Island, Victoria Land, lasted twenty-five minutes.

Days passed without further opportunity to land. The ice-bound coast, crested by mountains, was unyielding. At 76° south, due east of the South Magnetic Pole, they came to a lofty outcrop fortified by terraces full of birds. Ross named it Franklin Island. If they were not to reach the pole, Ross reckoned this might be their nearest landing to it. They set off again in the boats, this time with a Union Jack made of silk, gift of Anne Coulman. Typically, the island proved farther off than thought, and when they at last pulled near, much more formidable. Cliffs that seemed surmountable through the telescope now loomed three hundred feet above them.

They almost circled the island before they spotted a small sandy beach under a cliff on the southeast side. Ross crossed into the *Terror*'s whaleboat, the sturdier of the two, and threw a line back to the *Erebus*'s cutter. As the boat neared the beach, the breaking waves threatened to smash the boat. The cutter hauled them off just in time. When they came to a protruding dike, Ross stood up again, balancing himself on his second-in-command's shoulder. When Crozier suggested that putting his hand on the rock would certainly count as landing, and be sufficient for naming rights, Ross answered with a joke and instantly jumped at the ice-slicked rock as a wave crashed over him. Crozier barely followed, but two officers fell into the sea, clinging for life to the cutter's line. Then a third fell in, with no rope, and disappeared in the surf. By chance, he surfaced near the boat, pale and gasping. His messmates dragged him in, stripped him, and wrapped him up in their jackets.

Spectating helplessly from the rock, Ross came to his senses. Their rights of exploration in Terra Australis came with strict conditions, unlike any in the Arctic. The land was invitingly vast but offered no beachhead or gateway for exploration. The sea opened southward but the temperatures, perishing even in midsummer, made overwintering unthinkable. They were unwelcome guests here, on temporary permits. After Franklin Island, Ross forbade all further attempts at landing. The dip circle had pushed beyond 88°. The South Magnetic Pole was tantalizingly close. But for what remained of the Antarctic summer of 1841, their craving for discovery must be satisfied from the ship's deck.

Far from being sucked into a hollow Earth, as the Americans believed, the British ships found themselves repelled from the pole the closer they attempted to approach it. Days of storms and a strong current, both from the south, drove them backward the way they had come. Thick snow, with flakes like tiny stars, blanketed the decks and blinded them. Waves breaking across the ship froze as they fell, covering everything—deck, spars, faces—in layers of ice. The instant a man got wet, his clothes hardened to a frozen sheath. A rope drawn up from the sea was encased in ice two feet thick. Fantastic icicles hung from the rigging, some sharp as knives, others like strings of beads or bracelets. On deck, they kept a wary eye aloft for falling ice. Looking across at the *Terror*, they saw a mirror image of themselves: a ghost ship, frosted white from bow to stern, laboring in an icy surf.

Then the weather cleared, and the westward mountains came into view again, set against an intense blue sky. A dense field of ice lay between the ships and the coast, thirty miles distant. At midnight, the flattened sun skimmed the horizon, bathing the impenetrable pack in an eerie red light. Hooker's tow net brought up a variety of sponges, sea spiders a foot long, wriggling worms, and a cucumber-shaped invertebrate. One ice-loving mollusk wore a shell like chain mail; another, touched gently with the finger, glowed bright emerald. Ross recorded the ships' position daily, stuffed the paper in a bottle, and threw it over the side. If, by some chance, a bottle washed up on some distant shore, it would shed light on the direction of hemispheric currents. Also, if the *Erebus* and *Terror* never returned from the ice, it would establish their historic southing for the record.

On the morning of January 28, a fresh wonder appeared: an Antarctic volcano. The mountain, dead ahead in their southerly course, emerged from the mist shooting flame and steam. Eyes trained on the summit, they watched a dark smoke cloud belch out, followed by jets of fire. Heat had melted the ice at the peak while, farther down the slopes, the melting snow had refrozen into long, serpentine pools, mimicking the flow of lava. They glinted like metal in the clear sunlight. High up, massive columns of dark cloud wreathed the summit, with cirrus-like streams drifting at elevation toward the mountain range in the west. Still higher, the moon hung like a second sun in the blue sky.

Fig. 12.2. Mount Erebus, smoke issuing from its summit, depicted with the legendary exploring vessel for which it was named by James Ross.—Frontispiece, Joseph Hooker, *The Botany of the Antarctic Voyage* (1847). The Royal Botanic Gardens Library, Kew.

The British expedition included no poets or painters in its ranks, but many aboard were devout. They experienced the scene of the polar volcano as an almost religious vision. The pitch-dark volcanic cloud, streaked with flame, highlighted the sky, purer than any tropical blue. When the sun lowered, tints of gold and red played along the rim of the cloud, supercharging the effect. They had not reached the pole, but this fiery southern junction, where all first elements of Earth's creation—mountain, ice, sea, and sky, with the sun and moon in witness—seemed to merge in a single, stained-glass tableau, touched them to unspoken depths. An illiterate sailor from the *Erebus*, two years later in Rio, made a friend write down what he had seen that morning, like a visionary testament. The biblical scale of Antarctica awed his recollection. Ice islands separated from the barrier, by themselves, were "Large enough to build London on their Summit." Having reached the ends of the available Earth, James Ross named the volcano, and a quiescent companion peak just to the east, after the hardy ships that had brought them there.

A navy man to his core, James Ross cursed Terra Australis for obstructing further sailing progress. At 78° south, hundreds of miles beyond where humans had ever been, with the ice pack encroaching and winter coming on, he knew his Antarctic campaign was nearing its end. As he peered over the side in a calm, the young winter ice congealed on the water before his very eyes, generating its sinister, clicking sound. Soon it would seal the ships in. Barely a hundred miles from the magnetic pole, the compass spun uselessly. Ross's technological advantages, as a Royal Navy captain and emissary of human civilization, were being stripped away, leaving them more vulnerable by the day.

But Ross could not resist one last invitation to explore. Steering to the east of Mount Erebus and Mount Terror, a new breathtaking novelty presented itself: a great shelf of ice, two hundred feet high, perfectly flat, and stretching east to west as far as the eye could see. The four-year voyage of the *Erebus* and *Terror* involved the customary, months-long stretches of crushing boredom. But twice in two days, in January 1841, the Victorian explorers found themselves dumbfounded by the incomparable physical beauty of Antarctica. James Ross had covered the majority of the globe, while bookish Joseph Hooker had read every available illustrated volume of natural history. But neither could draw on any earthly analogy for what they were looking at: a giant seaward platform of ice, perfectly regular, without crack or fissure anywhere on its face, and seemingly without end. The great plug of the world.

Ross did not think to name this prodigious phenomenon after himself. To him, it represented, like all physical features of Antarctica, an obstacle to his cherished pole. So he called it the Great Ice Barrier. At first, they had no inkling of its true dimensions. Skirting eastward along the barrier that first evening, they hoped to clear it by midnight. Ross expected, every moment, to hear the lookout's cry of open water to the south. But the farther and faster they sailed, the more the shelf extended itself, in unremitting sameness, to the easterly horizon: a literal wall of ice that soared above them twice the height of the masthead. Fountains of spray, bright blue and green in the sun, showed where the waves pummeled the base of the shelf. Local icebergs, too, crowded with penguins, signaled the effects of glacial erosion. But to the evangelical company aboard the *Erebus*, the great barrier, taken whole, seemed immutable, like a statement made at Earth's genesis and unchanged since—a God-ordained limit to their voyaging speculations. As for what might lie beyond the barrier, it was almost blasphemy to imagine.

On the eighth morning of their barrier cruise, the explorers found themselves enclosed in a small bay hollowed out in the shelf, surrounded on three sides by dazzling parapets of ice. Then, by chance, the sun fell behind a cloud, softening the light. Though they had crossed the Antarctic Circle a month ago and skirted along the mysterious barrier for a week, this happenstance blend of ice, atmosphere, and color was entirely novel, as if this nameless bay with its strange halo of light were the destination Antarctica had in store for them all along. The officer on watch whispered an order to fetch the captain. When Ross emerged from below, he gazed up and around, blinking, as if adjusting to the unfamiliar aura. Then he stopped still, transfixed like the rest. More men appeared from below. They stood on deck in silence, while no orders were given. Time passed, and their limbs froze painfully in the wind off the ice; but they continued to stand there, bathed in polar light.

‡ 13 ‡
Homecoming

The penguin kidnapped from the newly baptized Dumoulin Islands off Adélie Land in early 1840 had battled through the last hours of its existence on the deck of the *Astrolabe*, as a source of merriment to the crew. It was given a vulgar nickname, then butchered and stuffed. In the captain's cabin, the expedition naturalists Jacques Hombron and Honoré Jacquinot made notes on the curious polar land bird, the southernmost animal on Earth. They noted the flecks of blue on its feather tips, and the white feathers growing more than halfway along its beak that disguised its true length (and fraternity with its giant forebears). The stuffed Adélie then made the first recorded penguin crossing of the equator, as Dumont D'Urville hurried home to announce the French triumph in Antarctica and promote his scientific collections. He was now a full two years ahead of his American and British rivals.

The *Astrolabe* and *Zélée* docked in Toulon on the morning of November 7, 1840. The local welcome was subdued, as so many of the ships' companies had not returned. The following summer, Hombron published his official description of the Adélie penguin in the *Annales des Sciences Naturelles*. Then, in spring 1842, the expedition's treasures were opened to an invited public at the Jardin des Plantes in Paris. Selections from the five thousand strong rock inventory were put on show, including granite and gneiss fragments from the Adélie Coast.

Among the fauna, D'Urville's Adélie penguin was star of the show, outdrawing the dugong, the long-nosed Asian monkey, and gaudy tropical parrots. The first European public to see the Adélie were enchanted by its anthropomorphic mien—part child, part soldier—launching the penguin's long and continuing career as a zoological crowd pleaser.

Alongside the stuffed Adélie, skulls of creatures from various latitudes were organized to show the importance of climate in promoting animal diversity. But it was a display of Polynesian plaster busts—courtesy of the expedition phrenologist Pierre Dumoutier—that challenged the Adélie penguin for Parisians' attention at the Jardin des Plantes. The impassive heads, with their supposedly unique indentations, promised to advance racial understanding of cultural differences D'Urville himself had already established, through groundbreaking linguistic scholarship, on earlier journeys in the Pacific.

D'Urville believed implicitly in the "science" of phrenology, which taught that a person's character and intelligence, even his destiny, was written in the bumps and fissures of his skull. As a navigator who had sailed three times around the world, it seemed common sense that one might know a man's mind from its topographical features, just as he would judge the character of a new land by careful tracing of its coast. Phrenology, to its adherents, offered the anthropological equivalent of a fossil dig. The human skull, like the bones of prehistoric animals unearthed by Baron Cuvier in the Paris Basin, could yield up profound secrets. Better still if the subject was still living and could pay for the information.

The phrenologist D'Urville had consulted before sailing for Antarctica did not come cheap, but his analysis had been worth every penny. For fifteen minutes, he pressed his fingers deep into D'Urville's unruly grey cone of hair to probe his scalp. He referred to the diagrammatic plaster skull on his desk. He took notes and consulted charts. Then he began to enumerate, in an uncannily precise list, all those qualities of character in which D'Urville himself took personal pride: his intelligence and courage, his persistence against adversity, his feeling of superiority to other men. He even alluded to some great destiny still awaiting him. D'Urville had been pleased, and Pierre Dumoutier was duly embarked as the first and last phrenological research scholar in the Southern Hemisphere. His celebrated collection from the French Antarctic Expedition remains in the collection of the Musée de l'Histoire Naturelle in Paris, where the heads are occasionally brought up from the basement for display, with all the necessary apologies for their acquisition.

Even in the darkest days of his final voyage, during the dysentery outbreak in late 1839, D'Urville did not forget the promise of phrenology. With groaning men on cots lining the deck of the *Astrolabe* a few feet above his head, D'Urville sat down at his desk to redraft his will. In the likely event of his imminent death, he directed that his body should be delivered into the sea, but not intact. His heart was to be extracted and given to his wife, and his head separated and preserved by Doctor Dumoutier as a gift to phrenology. D'Urville survived that time of crisis in the Indian Ocean, but Dumoutier, it turned out, played a significant cameo role in his commander's actual death rites, which came soon after.

The newly promoted D'Urville had orders to report to King Louis-Philippe at the palace immediately on his return from the Southern Ocean. As it was, it took the old explorer two months to comply. An old friend who paid a visit to his garden house in Toulon was shocked. The bear-like, imperious Admiral D'Urville was reduced to "a specter, a worn-out body." His third voyage south had broken him. Adéle, it seems, was little better. She had been a vivacious young woman, but the deaths of three children had installed a despondency not even her husband's long-awaited return could shift. Signs point to an opiate addiction. Adélie only belatedly joined D'Urville in Paris, when at last he made his appointed rounds to the king, the academicians, and his expectant publishers. In their apartment near the Luxembourg Gardens, husband and wife slept in separate rooms. D'Urville suffered chronic headaches, and Adélie from her long-term abdominal complaint. Among the doctors who attended them was Pierre Dumoutier, who had established himself as a family confidant.

D'Urville lived long enough to see the second volume of his voyage history appear in Parisian booksellers' windows. It included a lavish foldout map of the *Astrolabe* and *Zélée*'s first south polar campaign, accompanied by his gripping narrative of their hairsbreadth escape from the ice pack east of the Antarctic Peninsula in the summer of 1838. Ever the workaholic, he was immersed in writing the next volume when on May 8, 1842, he, Adélie, and their sole surviving son Jules took a vacation day to the palace grounds at Versailles for the king's birthday celebrations.

Late in the afternoon, after the spectacular fountain show, a huge crowd jostled for seats on the newly opened train back to Paris. Seventeen cars carried more than a thousand passengers, with the D'Urvilles seated near the front in the covered sections. As the crowded train left the station, witnesses observed the first locomotive lurch abruptly across the tracks. Half way to Paris, the axle broke clean off, sending the lead vehicle into an embankment. Amid screams and the ripping of metal, a six-car concertina followed. After a dreadful pause, the coal from the engine ignited, engulfing the wreck in flames. In the early days of rail, passengers were locked into their compartments, so the crash survivors had no chance of escape. For the desperate onlookers, kept at bay by the prodigious heat, it took fifteen minutes—an eternity—for the shrieking to die away. D'Urville, whom some of his fellow passengers had recognized, was last seen holding his sleeve to his mouth, yelling for someone to save his wife and son.

More than two hundred passengers died in the May 8 train disaster, the worst in Europe to that time. They laid out the carbonized bodies along the twisted track, and summoned Pierre Dumoutier. Like a fossil hunter on an Antarctic beach, the phrenologist searched among the remains for his commander. He identified Adélie from a locket and chain about her neck, and laid her beloved Jules beside her. Then Dumoutier bent down and picked up a charred skull from the still-smoking mess. He considered its size, made a show of inspecting its cranial dents, and pronounced it D'Urville's.

Husband, wife, and son are buried together in Montparnasse. The government-funded monument illustrates the explorer's great feats, climaxing in the planting of the French flag on Terre Adélie. But the raw irony of his death is left to the imagination. Dumont D'Urville, France's greatest navigator, had braved the frozen Antarctic and brought his ice-battered ships home in triumph, only to perish on a day trip outside Paris, trapped with joyriders on a burning commuter train.

Charles Wilkes's homecoming was more ignominious than tragic. Publicity for the epic American voyage, what little there was, focused on the bitter feud between the commander and his officers. Instead of celebrating their historic, fifteen-hundred-mile voyage mapping the East Antarctic coast, the ambitious lieutenants of the US squadron were distracted by hatred for the commander who, in their minds, had ruined their chance at glory with his woeful seamanship and petty malice. Their list of grievances, nursed over thousands of nautical miles, buried the mission's achievements.

Lieutenant Joseph Underwood from Rhode Island was exactly the type of young officer—educated, skilled, brave—to attract Wilkes's paranoid rage. He had hounded the uncomprehending lieutenant all across the Pacific. Then, during their tour of the Antarctic coast in the last week of January 1840, the simmering feud broke into an ugly public skirmish. Skirting the pack, the *Vincennes* had come upon a twenty-mile-wide bay littered with ice. Wilkes probed southward for a few hours among the giant, tabular bergs before he lost his nerve. Underwood, the officer on watch, saw the chance of polar fame slipping agonizingly away. He stopped short of confronting Wilkes but did the next best thing. He took up a piece of chalk and announced on the deck's public log that "a southward opening had been reported." Seeing the entry the next morning, Wilkes exploded. He turned the ship around and spent a full day retracing their journey to prove Underwood wrong. Back in the inlet, the treacherous pack obliged Captain Wilkes by canceling the lieutenant's southward passage from view.

Such was the demoralizing dispute aboard the *Vincennes* at "Disappointment Bay" in summer 1840—a scene that would surely have been relitigated at Charles Wilkes's post-expedition court martial, had Underwood lived to see it. Only months after surviving the storm that brought the Americans' Antarctic mission to its end, the lieutenant was clubbed to death by Fijians on the beach at Malolo. As if rehearsing the old quarrel between captain and officer, historians can't agree on the precise location of Disappointment Bay. Baffled by the dustless glare and horizon-bending mirages, Wilkes was full degrees of latitude farther from the Antarctic coast than he imagined—a mistake for which he paid dearly in the expedition's aftermath.

In the late summer of 1841, the *Erebus* and *Terror* were making a northward return from their historic season of exploration south of the ice pack. Despite feats of his own to report, including a record southing, James Ross detoured west in the vicinity of 70° south to make a critical assessment of his American rival's claims. On his cabin desk lay a taunting newspaper clipping from the *Sydney Herald*, in which Charles Wilkes had announced his "Discovery of an Antarctic Continent" for America. Ross's melancholy object in this extended cruise was to confirm, for himself, his own worst fears: that he had been beaten to the prize.

On the morning of March 4, the *Erebus* had drifted into an apparent bight, rimmed with glaciers, when the breeze suddenly freshened to a gale and set the ship pitching so that her bowsprit smashed hard onto the ice with every plunge. An icebound coast, perhaps islands, appeared in the west but was soon lost to view in the thick snow. Across the way, the storm had ripped several of *Terror's* sails to shreds, so Ross ordered both vessels to hove to for repairs.

The delay allowed Ross the opportunity to consult, once again, a most extraordinary map in his possession. It indicated coastline extending sixty miles from the southwest to the *Erebus's* immediate vicinity—a dangerous lee shore. The map had been sent to him, the previous April in New Zealand, by none other than Charles Wilkes. An accompanying letter rejoiced in the American successes in East Antarctica. This despite orders strictly forbidding the American commander from broadcasting his discoveries. Wilkes, however, basking in his triumph, was unable to resist a personal message to Ross, telling him, in essence, that the race was over. Enraged at Wilkes's gloating communication, Ross could only stew in silence. The map was an insult he could not publicly resent because it had been offered to him in the guise of Anglo-American friendship, as a helping hand from one naval man to another.

For James Ross, however, trapped in a strange bay with a storm coming on, Wilkes's chart was a potential lifesaver.

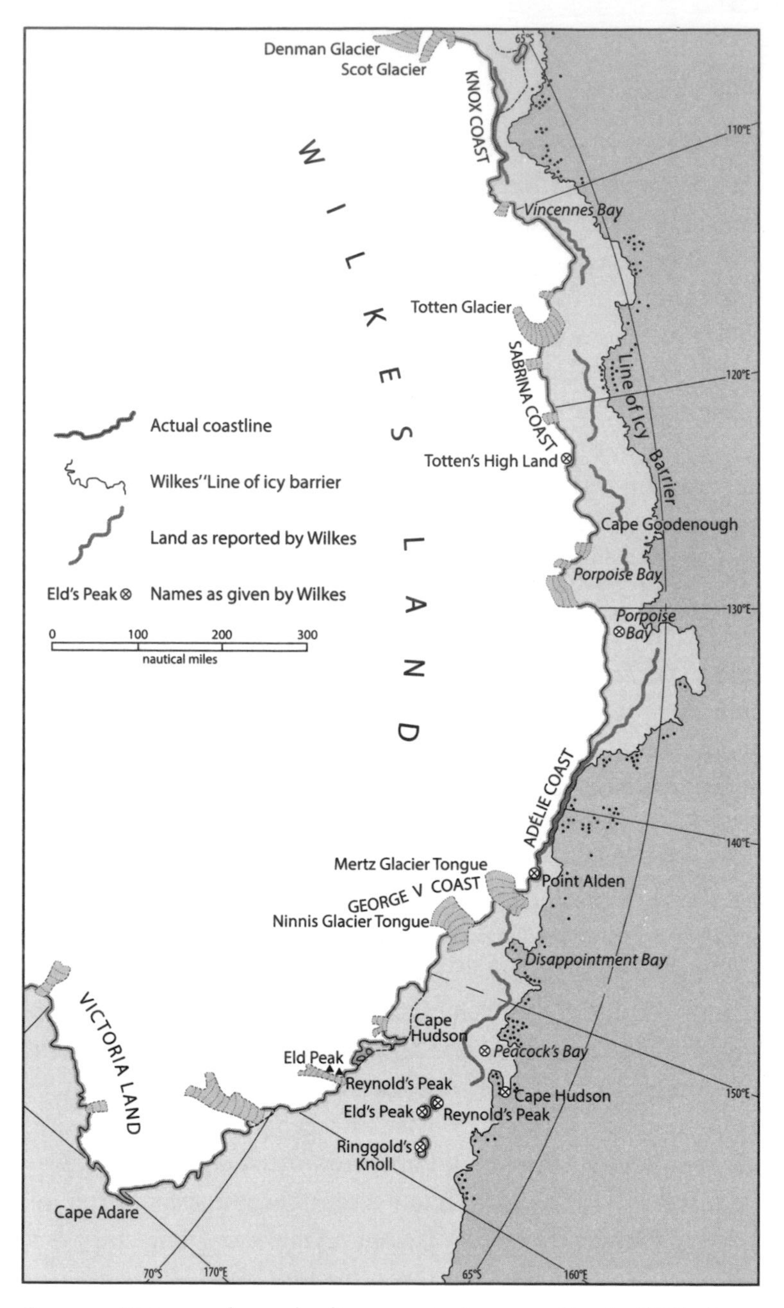

Fig. 13.1. This map shows the discrepancies between Charles Wilkes's notorious 1840 chart, which he sent to James Ross, and the actual coastline. The reasons for Wilkes's mistakes, and the extent to which they diminish his achievements in Antarctica, are still matters of international dispute.

During the brief night, he pressed as much sail as he dared to escape the snare, even as the continual parade of icebergs through the whipping snow forced the *Erebus* to tack toward the shore shown on Wilkes's map. Under the rug of darkness, with the cry of penguins all about them, everyone onboard was convinced of the looming presence of land. While the wind howled about their ears, they listened intently for the fatal sound of waves against rocks, or the jolt of the hull grounding on unseen shallows. Nerves in shreds, the British sailors prayed for survival, waiting for first light to reveal the full gravity of their predicament.

Imagine the feelings aboard the *Erebus* when the impatient sun revealed a pure vista of open ocean glinting with Mediterranean purity as far as the eye could see. Ross ordered soundings. The watch officer shook his head: no seafloor above six hundred fathoms. He hurried to his cabin with a mixed rush of feelings—bafflement trending toward glee. The happy truth soon dawned. The *Erebus* and *Terror* had spent the night in imaginary peril, navigating a treacherous shore that did not exist. Now, under blue skies, they were cruising gaily *over* Wilkes's mountain range. They sailed in circles all that day and the next, following Wilkes's chart, but the Americans' Antarctic continent was nowhere to be seen. Now, when Ross contemplated Charles Wilkes's map, it was not with fear and loathing but gratitude. With this botched chart, Wilkes had handed him a devastating weapon, and a late chance to snatch Antarctic glory for himself and Britain.

On his return to Australia, Ross reported Wilkes's fictitious Antarctic coastline to the Admiralty and composed a biting article for the *Sydney Herald*. Back in the Bay of Islands in New Zealand, he met with a supposed friend of Wilkes, named Captain Aulick. They spent an evening together aboard the *Erebus* comparing Wilkes chart with Ross's own. They shook their heads at the American's bold line of phantom coast. Perhaps Ross hoped that the intervention of a sympathetic compatriot would persuade Wilkes to quietly withdraw his claims. But Aulick belonged to the legion of American officers who believed command of the US Exploring Expedition should have been theirs by right and hated the upstart lieutenant from New York. The disgruntled American took the story of his meeting with Ross, and his firsthand view of Wilkes's incriminating chart, to the Sandwich Islands and Honolulu, where local newspapermen seized on the scandal.

The Wilkes Land controversy continued to simmer for a full century, in the form of a transatlantic paper war conducted at ever more rarefied levels. In the controversial aftermath of their Antarctic race, Wilkes blamed Ross for his court-martial and attacked him in print. A generation later, by way of retaliation, Clement Markham, the doyen of Victorian polar historians, systematically erased Wilkes's name from all Antarctic maps. Then, in 1904, Robert Scott seemed to vindicate Ross by sailing the *Discovery* directly over Wilkes Land, as the *Erebus* had done decades earlier.

But the battle over Charles Wilkes's legacy would not die. In 1914, Douglas Mawson—the first to land and explore the Wilkes Land coast itself—offered a qualified endorsement of his maligned predecessor. Some features on the American's 1840 map were correct, others not. The brilliant, dust-free air had deceived the navigators aboard the *Vincennes*, making the land appear closer than it was—a mistake he himself had made in the same waters. Regardless of errors of latitude, Antarctica was indeed a continent, and Wilkes thereby vindicated, in the main.

This was all that a confident new generation of American historians needed to take up their beleaguered countryman's cause. The humiliating circumstances of Wilkes's 1842 return had sunk the US Exploring Expedition from public notice. Even his obituary made no mention of him as the "Discoverer of Antarctica." So, they dug up old German maps from the 1840s—the only ones to recognize Wilkes's chart—and lobbied hard for a do-over of the first race to Antarctica. The results reflected the altered balance of power since Victorian times: the British conceded defeat, the French clung to a sliver of territory—"Adélie Land" was downgraded to "Adélie Coast"—while the Americans took the lion's share. On modern maps, "Wilkes Land" honors the US Ex. Ex.'s survey of the fifteen-hundred-mile stretch of coast between 100° east and 142° east. Only the territory east of Point Alden—where Ross cruised blithely over Wilkes's mountains—reflects the lingering trace of transatlantic bad feeling over who-did-what-to-whom in 1840 and afterward. A British king commands the surface territory—as George V Land—while Charles Wilkes, suitably, lurks below, in the form of the glacially unstable Wilkes Basin.

President Martin Van Buren triumphantly announced the Americans' discovery of Antarctica in his State of the Union address in 1840, when Charles Wilkes was hero of the hour. But by the time the *Vincennes* dropped anchor in New York harbor in June 1842—the last of the great sailing expeditions to circumnavigate the globe—his dispute with James Ross over the notorious map had gotten abroad. To compound his public relations nightmare, Wilkes was then accused at a court-martial brought by his officers with *never having seen Antarctica at all* on the discovery date in question. He had altered the *Peacock*'s log to cheat the French of their claim, and brought shame on the United States. Though Wilkes angrily repudiated the charge for the rest of his life, his reputation, with that of the US Ex. Ex., never recovered.

Instead of a hero's welcome on his return, Wilkes was subjected to humiliating days in the dock, listening to his officers give vent to four years' worth of bottled resentment. In front of a grim-faced panel of admirals and a salivating press gallery, the American commander was charged with excessive use of the lash, murdering natives in Fiji, falsely claiming the rank of captain, and bad language—in addition to lying about Antarctica. After two weeks of testimony, in which the worst of human nature was both narrated and displayed, only the count of illegal flogging stuck, for which Wilkes's punishment was a reprimand delivered on the spot in a few short sentences.

Herman Melville, for one, thought Wilkes deserved more, even a novel's worth. Their fathers had been prominent New York businessmen and moved in similar circles. Both sons were veterans of the South Seas, and Melville borrowed passages from Wilkes's multivolume expedition *Narrative* (published in 1845) for his Pacific seafaring novel *Typee*. Wilkes's *Narrative* thus occupied a prominent place in Melville's library (and imagination) while he wrote *Moby-Dick*, whose very title was inspired by another US Ex. Ex.–related publication: "Mocha-Dick; or the White Whale of the Pacific," written by none other than Jeremiah Reynolds, the original promoter of the American Antarctic mission and a longtime nemesis of Charles Wilkes. Beyond that, the literary evidence is circumstantial, but the lurid tale of Wilkes's trial in the New York newspapers in 1842, followed by his vengeful, paranoid prose in the voluminous *Narrative*, offered Melville a compelling case study in the pathologies of command—a real-life Captain Ahab.

The Stormy Petrel's list of enemies was long, and he kept adding to it to the end. After the notorious *Trent* Affair, in which Wilkes almost succeeded in goading Britain into the US Civil War on the Confederate side, Secretary of War Gideon Welles court-martialed him again—this time for disobeying orders and violating the neutrality of foreign ports for personal gain. Wilkes was pardoned by President Lincoln.

Dismissed from service, he spent his last years settling scores in an unpublished memoir totaling hundreds of pages of angry scribble—a fitting last testament from a towering misanthropist of the Victorian Age.

As for the rows of specimen jars in James Ross's cabin aboard the *Erebus*—filled with enough novel sea creatures to revolutionize Victorian science—their contents were never published. When Ross died in 1862, the Admiralty sent Joseph Hooker—now a pillar of the British scientific establishment, and destined to be the last man alive to have walked the decks of the *Erebus*—to Ross's home in Buckinghamshire to recover the collection they insisted belonged to them.

Hooker took the new train line from Kew to Aston Abbots, where the drawing room window overlooked a small lake with miniature twin islands Ross had named Erebus and Terror. Out the back, behind the kitchen garden, Hooker found a dirty rubbish pile of broken jars. It was all that remained of the collection he and Ross had labored over on the Antarctic expedition two decades before. Hooker felt a wave of ancient resentment toward his captain, who had taken advantage of his labors and arrogated a right to scientific recognition he then never exercised. The loss to science was immense.

In time, however, Hooker's anger abated, to be replaced by more sympathetic feelings toward his old captain. At first, on their return, all had gone well for James Ross. He retired from the sea, married his beloved Anne, and had four fine children. They were, it was said, a blissfully happy couple. But in the world beyond Aston Abbots, Ross's fortunes slid. His two-volume memoir of the Antarctic expedition had not sold well. He had no gift for dramatic writing, and publication coincided with the departure of his friend John Franklin to the Arctic on a lavish expedition to discover, at last, a route through the Northwest

Passage to China. Franklin had taken the *Erebus* and *Terror* and many of his friends with him, including Ross's second-in-command in the Antarctic, Francis Crozier.

Then, unaccountably, Franklin and his ships disappeared. For twelve long years, the search dragged on. Ross broke his promise to Anne and twice left Aston Abbots to return to the Arctic in search of his friend and his old ships—but in vain. It was on Ross's advice that the Admiralty directed its search to the north when, as it turned out, the expedition had come to grief heading south for home. Franklin had died early, leaving Crozier to abandon the ships and lead a dwindling band of doomed survivors across the ice.

In the midst of the turmoil over Franklin, Anne took ill and died. This was the blow from which Ross never recovered. He tried keeping up appearances for the sake of the children, but the grief was too much. He turned to drink and died alone. News of Ross's death was lost in the national hysteria over John Franklin. Ross, Parry, and the rest of the first Arctic generation were all forgotten in the ritual elevation of Franklin as the great polar martyr of the age. When friends of Ross petitioned the government to raise a public monument to the discoverer of the Antarctic continent and the North Magnetic Pole, they were refused. No one could be allowed to subtract from the glory of Sir John Franklin, the tubby underachiever who had led one hundred twenty-eight men to their deaths.

The mystery of the lost ships *Erebus* and *Terror* lasted a century and a half. Then, within two years—the summers of 2014 and 2016—search teams miraculously discovered them, sixty miles apart, resting quietly intact on the sandy floor of the Canadian Arctic. The circumstances of the final journey of these storied vessels—whose greatest discoveries were recorded at the opposite end of the world, under the command of James Ross—will never be known.

Better then to remember the *Erebus* and *Terror* as they were in the port of Hobart in their Victorian heyday. One unforgettable night in 1841, the ships were tethered together, decorated with lanterns and flowers, and their decks cleared to serve as a giant outdoor ballroom. Governor Franklin and his lady were throwing a grand party to celebrate the triumphant return of Captain Ross and his men from their adventures in the south, where they had solved at last, for England and all mankind, the mystery of Terra Australis. Though they didn't know it, the last three captains of the ill-fated *Erebus* were assembled aboard for the last time that night, when James Ross, John Franklin, and Frances Crozier—together with Joseph Hooker and Lady Jane Franklin—waltzed together in happy unison on her decks.

POSTLUDE

⚓

Last Ice

The first race to the South Pole resulted in a split decision. The French were the first to sight Antarctica and make landing, while the Americans charted the greatest stretch of coast and established its continental dimensions. The British, meanwhile, who were the last on the scene, traveled the farthest and saw the most. James Ross's extraordinary 1841 voyage blazed the trail for polar explorers of the later Heroic Age by whom he, Wilkes, and D'Urville have been unceremoniously eclipsed.

The Victorian explorers' greatest legacy, however, lay not in national conquest, but in their shared commitment to scientific discovery. These were the first humans to properly encounter the seventh continent's icy grandeur, and their mixed feelings of wonder and terror mirror our own growing awareness of an existential threat issuing from the polar south. Even as they struggled for survival in the icy seas—conditions for which they were suicidally ill-equipped—they doggedly mapped, recorded, sketched, and sampled all they met with, however confounding. Nearly two centuries later, the business of Antarctic data collection is an empire unto itself, a vast domain. Though the Victorians retreated in awe from the ice continent, stymied in their efforts to make landing and claim the pole, they are its true founders as an object of knowledge.

The Victorian discoverers live on, likewise, in a suite of place names increasingly familiar to our own age of polar obsession: the magnificent Ross Sea and Ice Shelf in West Antarctica; Wilkes Land and its Basin in the east. Dumont D'Urville's name, unfortunately, is attached only to a French research station, a negligible island and peak off the Antarctic Peninsula and, on some maps, a patch of sea. But long-suffering Adélie D'Urville has her penguin, which, though now threatened by a deteriorating Antarctic habitat, is a proven climate-change survivor and might yet find a way to outlive us all.

In this era of rapid glacial melt at the poles, the bitter Anglo-American dispute in the 1840s and beyond over the exact contours of East Antarctic coastline can seem like a luxury. But to its Victorian discoverers, land was the prime currency of exploration. As emissaries of agricultural nations, ice was a "barrier" to their true object. Ross, D'Urville, and Wilkes peered across the vast white belt surrounding the mystery continent, desperate for dirt and terrain, for an ecology they could recognize. By these standards, for Charles Wilkes to mistake an ice shelf for mountainous coast was a cardinal error. But that is all changed now. The scientific descendants of the Victorian explorers mostly bypass the land except as a support infrastructure for trillions of tons of ice. Because it is the nature of Antarctica's glaciers, not the land beneath them, that will determine the sorts of lives we, the global human community, will lead in the coming centuries.

In 2010, Leg 318 of the Integrated Ocean Drilling Program (IODP) set out to better understand the waxing and waning of the East Antarctic ice cap since its initial glaciation thirty-four million years ago. In particular, how had the ice cap responded to global warming episodes of the past—as in the Miocene and late Pliocene epochs, when Earth's temperatures rose to levels we are now rapidly approaching in the twenty-first century? Given East Antarctica's land elevation—it is the most mountainous territory on Earth—the prevailing assumption was its massive ice sheet was safe from anthropogenic warming. But with almost two hundred feet of potential sea-level rise locked in the East Antarctic ice sheet, it paid to be sure.

So, the IODP's *Glomar Challenger* set sail from the New Zealand port of Wellington, bringing two supplementary passengers: a polar meteorologist and an ice watcher with deep Southern Ocean experience. Both were kept busy during the two-month voyage, as the *Glomar Challenger* traversed the treacherous seas first charted by D'Urville, Wilkes, and Ross one hundred eighty years ago. Their object was the still-unexplored continental margin, a dark undersea world of alternating troughs and ridges, transected by canyons, and with a precarious upward slope extending landward, beyond their reach, toward the recessive base of the ice shelf. Unlike the land-hungry Victorians, the terrain that mattered to the IODP scientists lay beneath the ocean floor, in marine fossil records of ice sheet fluctuation buried in a five-mile-deep sedimentary column. No one doubted the existence or importance of this submarine treasure trove; the challenge lay in retrieving it under the worst maritime conditions in the world.

The research ship was not halfway to its first destination off the George V Coast when it ran into a low-pressure maelstrom delivering sixty-knot winds and forty-foot waves. When after three miserable days the storm lifted, the researchers saw their first penguins clipping the swell, and then icebergs en masse. The Leg 318 operation plan nominated seven drill sites, but the ice pack stopped them almost twenty miles short of the first, so they chose another to the northwest. Site 1355 was clear of ice, but technical problems plagued the drilling, and the cores drew only coarse sand and gravel. At the next site, they drilled a half mile deep to the Eocene-Oligocene boundary, only for an advancing storm to force them to abandon the hole.

With the northern drill sites exhausted, the *Challenger* had no choice but to turn south and risk the ice pack. For the next five weeks, at five separate sites, the drilling team dodged icebergs and whales, survived entrapment by the pack, rode out near-hurricane-force winds, and endured windchill temperatures of twenty below, which blanketed the decks and equipment in ice. On one occasion, a killer iceberg floated directly over a drill hole they barely had time to vacate. But it was all worth it. Sealed within their hard-won cores, and laid out in gnomic grandeur back on the dock at Hobart, was a vital message sent from deep time: a tale of Antarctic ice, stretching back fifty million years, from Hothouse to Icehouse.

The storm-battered Leg 318 mission of 2010 has produced a flood of breakthrough scientific papers: on Southern Ocean paleoclimate, glacial melt, sea-level rise, and the ever-changing relation between Antarctic land and ice. Charles Wilkes, were he alive, would have taken comfort in the bottom line: his fanciful 1840 map of the East Antarctic coast, or any map, was no more than a snapshot of a continent whose larger, dynamic existence over time is governed by tectonic plates, deep ocean currents, and Earth's orbit around the sun, among other planetary forces humbling to contemplate. From the epic data dive of Leg 318 emerged one clear signal relevant to our present circumstances, and with it an ominous new catchphrase in the climate-change lexicon: "marine ice sheet instability."

For analogies to our current era of global warming, the Leg 318 scientists looked beyond the world-chilling EOT, thirty-four million years ago (myr), to two more recent warm periods: the mid-Miocene (17–13 myr) and the mid-Pliocene (5–3 myr). Both warming episodes involved atmospheric carbon levels ranging between four hundred and six hundred parts per million, equivalent to our current levels and those predicted for 2100. By the mid-Miocene, the East Antarctic coastline bore no resemblance to the tropical diorama of the Early Eocene. Over the preceding twenty million years, its inland rivers had broadened into majestic fjords, like present-day Norway or Greenland, delivering seasonal ice to the cooling ocean. This peaceful rhythm was interrupted on occasion by dam-bursting ice-floods on a scale never seen in the human record. Then, as temperatures dropped further, the land became fully encased beneath its glacial canopy. The urgent question for Leg 318: how had elevated carbon levels in the past affected the great East Antarctic ice sheet? Was it immune to melting, as some models insisted, or had it been catastrophically undermined? Had an unstable Wilkes Land shed its frozen bulk, flooded the oceans, and redrawn the world's coastlines, as it might yet do again?

Attention focused, in particular, on the so-called Achilles' heel of East Antarctica, the glaciated coast fronting the Wilkes Subglacial Basin (WSB) that Lieutenant Joseph Underwood had been so eager to explore in the summer of 1840. The WSB is a massive, mile-deep crustal depression extending from the hinterland of the Transantarctic Mountains to the George V Coast, where its seaward outlet ranges a full four hundred miles in the neighborhood of the Ninnis and Mertz glacial tongues. While the majority of the East Antarctic ice cap drapes serenely over alpine ranges or plateaus—an unbreachable kingdom of ice—the glaciers of the WSB are grounded well below sea level, up to a half mile.

Emplaced so, the glaciers of this coast, like their counterparts in low-lying West Antarctica, are naturally vulnerable to changes in ocean temperature and circulation. But how vulnerable, and over what time frames? The WSB alone contains volumes of ice sufficient to raise global sea levels by thirty feet, enough to submerge the world's great coastal cities and send millions of refugees scrambling for higher ground. During the Pliocene Climatic Optimum—when CO_2 levels were last at the levels predicted for 2100—sea levels were *seventy feet higher* than today, implying significant ice sheet collapse in East Antarctica.

In her lab at the University of London, Leg 318 scientist Carys Cook scrutinized one marine core in particular, drawn off the Adélie Coast in Wilkes Land. Two hundred fifty feet in total length, the Site 1361 core offered a continuous geoclimatic history of the Pliocene, the intermittently warm epoch initiated five million years ago. What she found was something like Pharoah's dream in the Bible, with its fat cows followed by lean cows, signifying rich harvests and poor. In the 1361 core, Cook identified eight diatom-rich layers of submarine clay interspersed with eight diatom-poor layers. The diatom-rich portions represented periods of reduced ice on the Wilkes Land coast, when warm, biologically productive seas bathed the Antarctic shore. The diatom-poor layers, conversely, signaled a return to impoverishing cold.

Sifting through the warm-climate sections of the core, Carys Cook came across microscopic grains of igneous rock dating from the Jurassic origins of the East Antarctic crust. These grains bore no relation to the surrounding sediment and could only have been imported. From this point, Cook's chain of reasoning was simple and irresistible. The nearest igneous source lay hundreds of miles away in Victoria Land, deep within the Wilkes Subglacial Basin. The sole means by which these rock fragments could have been eroded was by relentless grinding at a glacial margin. Therefore, the Wilkes Land glaciers, which currently line the coast, or extrude tongue-like into the ocean, had retreated hundreds of miles from these maxima during the Pliocene, before rebounding and retreating again in a spectacular binge/purge cycle of glacial excess. The marine-based glaciers of Wilkes Land, like Ninnis and Mertz, were dangerously unstable and had in the past disappeared hundreds of miles inland under climate conditions similar to our own.

For an all-important second line of evidence to support this alarming conclusion, Cook turned to the results of an earlier IODP expedition, Leg 188, which hinted at a massive Pliocene ice-rafting episode in Prydz Bay, a thousand miles to the west of Wilkes Land. On the seafloor at Prydz Bay, scour marks a half mile deep record the transit of icebergs released from an epic ice sheet disintegration. The source of the ice-rafted debris at Prydz Bay could be nowhere else than the WSB, which had shrugged off its glacial cover in rising Pliocene temperatures like an unwanted blanket. An "iceberg armada," deposited into the ocean at a rate up to nine times faster than in our current interglacial period, included bergs fifteen hundred feet thick, dwarfing their present-day counterparts. The "massive iceberg production event" discovered at Prydz Bay, concluded Cook, was the natural consequence of a Wilkes Land glacial regime highly sensitive to even modest changes in temperature.

The tipping point for East Antarctic sea surface temperatures, according to the Leg 318 cores, is 3°C. Today, the freezing waters of the Wilkes Land continental shelf continue to enjoy the protection of katabatic winds and a powerful cold stream at abyssal depths. The circumpolar current is warming, however, and will soon cross the threshold at which the glacial massifs begin their irreversible self-destruction. During the Pliocene Climatic Optimum four and a half million years ago, the dominant westerly winds of the Southern Ocean shifted southward, driving oxygenated water masses downward along the East Antarctic margin and eroding the glacial shelf. Other studies show that with circumpolar seas at 3°C, the Antarctic ice cap becomes susceptible to orbital drivers (mirroring ice age rhythms in the Northern Hemisphere) and to small but persistent increases in atmospheric CO_2 and mean temperature—a world like our own, except with sea levels eighty feet higher than today. In short, we are currently charting a new course for humanity's future, sailing headlong back to the Pliocene.

A 2014 paper in *Nature* narrowed the implications of Wilkes Land marine ice sheet instability to the starkest possible formula. Matthias Mengel and Anders Levermann, modelers from the Potsdam Institute for Climate Impact Research, concluded that the warm-water erosion of a relatively small barrier or "ice plug" at the coastal margin would unloose the pent-up Wilkes Land glaciers on an irreversible cascade. Though this extreme scenario is not due to unfold this century, for the human generations alive to bear witness, it will be the Pliocene iceberg armadas all over again. When that time arrives, all the maps of our world—not just Charles Wilkes's—will be reduced to historical curiosities only.

Marine ice-sheet instability is a more immediate threat in low-lying West Antarctica, which contains ice mass equivalent to the entirety of Greenland. Changing global sea levels over the last several million years have been dominated by the advance and retreat of West Antarctic glaciers along their marine basins, suggesting the future will mirror the past.

Since 1980, warming waters in the faraway tropics have altered southern atmospheric circulation, channeling warm air over West Antarctica. These winds, in turn, have driven the circulation of so-called Circumpolar Deep Water (CDW) southward. The CDW is salty and warm, up to 4°C, and travels at depth along troughs in the seafloor. The CDW's new frontier, the Amundsen and Bellingshausen Seas, present a suite of ice shelves vulnerable to warm-water infusion, most notably those

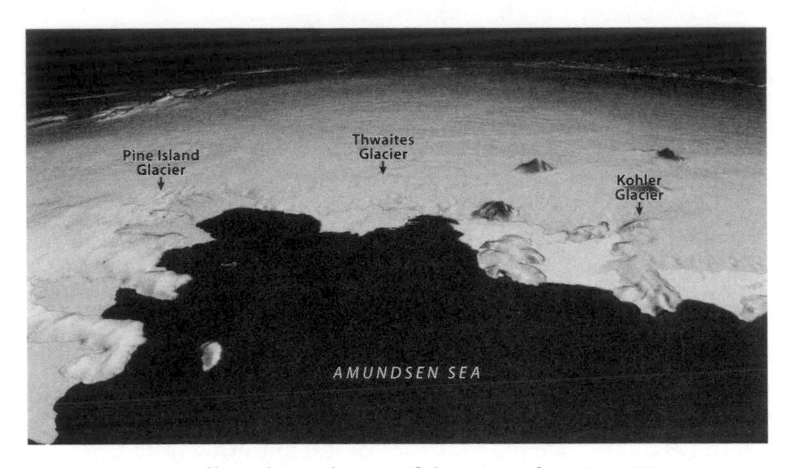

Fig. P.1. The rapidly melting glaciers of the Amundsen Sea, West Antarctica.—NASA/GSFC/SVS.

fronting the giant marine-based Pine Island and Thwaites Glaciers.

In two decades after 1992, the Pine Island Glacier retreated thirty kilometers. Ocean heat influx to the shelf face spiked, and the flow speed of glacial ice toward the coast doubled, launching icebergs into the sea in a slow-motion frenzy. Because of the reverse slope beneath it, this accelerated melting produces a feedback loop. The glacier's base now rests on a bed four hundred meters deeper beneath sea level than in 1992, exposing an ever greater area of its face to further erosion by the warming sea.

Pine Island's near neighbor, Thwaites—dubbed "The Doomsday Glacier" by *Rolling Stone*—is more concerning still. The size of Florida, Thwaites alone contains two feet of potential sea-level rise, enough to submerge Miami and much else besides. The discovery in 2018 of a miles-wide cavity in the glacier, the source of fourteen billion tons of ice now loosed to the oceans, suggests Thwaites will collapse this century. A team of no fewer than sixty researchers has now descended on the Amundsen Sea to track its decline. The International Thwaites Glacier Collaboration is the largest single research venture in Antarctica in decades.

In a runaway melt scenario, loss of these glaciers would irreversibly undermine the entire West Antarctic ice sheet, exposing the keystone Ronne Ice Shelf in the Weddell Sea to the east and the Ross Ice Shelf to the west. Surface melt from above-freezing air temperatures creates fissures in the shelf, while the warmer ocean flow beneath chips away at the base. Over time, glacial streams expand and the entire sheet drains into the surrounding oceans. In the future, trans-Antarctic seaways connecting the Amundsen, Weddell, and Ross Seas, last open in the Pliocene, will make for far friendlier sailing opportunities than the explorers of 1840–41 enjoyed. There will be an open polar sea at last, as the Renaissance geographers dreamed of. At the Ross Shelf, seismic sensors drilled in the ice have recorded startling vibration changes as its surface softens and fractures open in the glacial interior. The Ross Ice Shelf is "singing" its own requiem.

Historical precedents for this Antarctic end game are clearly identifiable in the sedimentary record. In the Pliocene, when Earth's climate conditions closely matched our own, the Ross Ice Shelf melted away, allowing the massive load of land ice behind it to drain away into the ocean, drowning coastlines worldwide. But the conclusion of the last ice age fourteen thousand years ago offers a much more recent example of the West Antarctic Ice Sheet's susceptibility to warming temperatures. Coral records from Barbados and Tahiti indicate that global sea levels rose at the rate of five meters per century to a high point of twenty meters (sixty-five feet) in less than five hundred years. A significant portion of that rise can be traced to West Antarctica. According to one 2016 model, the present sheet, under increased stress from warming, is due for a repeat collapse by 2250.

When it comes to predicting sea-level rise for our century, the Antarctic ice sheets are the joker in the pack. Pegged to temperature increases, seas might rise three feet by 2100 or twice that much. Either way, the costs of rising seas will top a trillion dollars annually by midcentury. Millions of people from low-lying coastal cities—from New York to Alexandria, Shanghai to Mumbai—will be forced to pack up and leave, joining a global exodus of up to two hundred million climate refugees worldwide. The humanitarian disaster set in train by the collapse of the Antarctic ice sheet will dwarf all forced mass migrations of the past, including the epic trauma of the Middle Passage. As coastlines are redrawn, the human social contract will be hastily rewritten, under emergency conditions not friendly to democratic process or human rights.

Beyond 2100, sea-level rise will not abate but only accelerate as more Antarctica glaciers reach their tipping point and more urbanized coastlines are inundated. This is without the prodigious interior ice sheets of East Antarctica, at their frigid elevations, contributing a single drop to the rising tides. But the great eastern ice cap is not invulnerable. Burning all available fossil fuels still in the ground would raise global temperatures sufficient to melt it completely, ensuring an eventual rise in sea levels of over two hundred feet.

The human inheritors of this disaster, centuries from now, will inhabit unrecognizable shrunken continents hosting devastated ecosystems. As for our imagining the lives our descendants will lead, perhaps the Patagonians D'Urville encountered in 1838 offer a clue. These resilient climate warriors in Tierra del Fuego mostly prospered during their long migration from the north but were trapped by slingshot climate change millennia ago: first by the so-called Antarctic Cold Reversal, then by melting glaciers. Their lives were nasty, brutish, and short; then they died out. Given these suggestive precedents, the current prospect of Antarctic melting sets the stage for an epic reversal in human fortunes.

In January 2017, my Antarctic touring ship sailed southwest past Cape Flying Fish, where William Walker almost came to grief in the pack in 1839, and into the remote Amundsen Sea. No ships of trade travel this route, nor is it popular among tourists. It is simply too far from anywhere. We approached Peter the First Island, a misty snow dome rising out of the pack, where fewer people have stood than on the Moon. But the surrounding ice was too dense for a landing.

The impregnable pack likewise kept us at a distance from the West Antarctic coast and its low-lying glaciers. But the dynamic changes underway along this vital coast were evident even to the passing polar tourist. One afternoon, under a gleaming blue sky, we fell among a flotilla of stadium-sized icebergs. Their perfect flat tops showed they had only recently been calved from the Pine Island and Thwaites Glaciers to the south. Iceberg production in these waters has increased 75 percent over recent decades, thanks to warmer sea and air temperatures. Iceberg armadas during the Pliocene deglaciation of Antarctica patrolled these coasts, when Terra Australis threw off its icy mantle and seas rose by hundreds of feet. Our ship felt suddenly tiny, as it literally dodged the bergs. We were like scouts sent to gauge the strength of an enemy—the frontline of a new iceberg armada.

A string of endless days later, we arrived at the eastern edge of the Ross Ice Shelf. Setting out in rubber dinghies, we explored the Bay of Whales, the southernmost waters on Earth, from where Roald Amundsen set out across the shelf on his sprint to the South Pole in 1912. He famously beat Scott, who, loyal to the memory of his Victorian predecessors, had launched his ill-fated bid from the other, less practical side of the Ross Shelf under the loom of the Erebus and Terror volcanoes.

For the Antarctic visitor, baffled by the scale and sameness of the alpine coast for mile after mile, the iconic Mount Erebus, with all its explorer legends, at least offers the sense of destination, of *having arrived somewhere.* Major scientific research stations—American, New Zealander, and Italian—are clustered nearby, human outposts on an inhuman terrain. Just as for Ross and his men in 1841, our sight of Erebus and the adjacent "Great Ice Barrier"—capturable within a single photo frame—inspired feelings of pilgrimage. The helicopters were revved up, and I squeezed in beside a Scottish doctor and a fireman from Seattle. Flying south over the snow-blown Ross Ice Shelf, flat as a Midwestern prairie, we were amazed to come to a channel of open water, sparkling in the sunshine. A giant slab of the seaward-fronting shelf, the size of a city, had separated itself from the main body and was floating out to sea. This glaciological wonder had not been visible from the ship. Antarctic ice shelves routinely calve their excess ice, responding to the pressure of the glacial flow behind them. But multiple times in the past, with temperatures only a few degrees warmer than today, the entire Ross Ice Shelf zooming by beneath us had broken up and drifted into the ocean, unloosing the vast reserves of land ice behind it.

Our pilot had a reputation as a cowboy, and he didn't disappoint. We flew breath-catchingly low through the channel, almost skimming the pale blue water as the ice cliffs blurred past on either side. We whooped and grinned at each other. In the pure thrill of being on the polar ice, it was difficult to focus on the distant Pliocene or any message Antarctica might be sending. On a months-long tour of the most unwelcoming place on Earth, actual pleasure is hard to come by. So we made doubly sure to enjoy ourselves on this once-in-a-lifetime flight, hotdogging through cracks in an ice shelf—as if we hadn't a care in the world.

ACKNOWLEDGMENTS

My deep thanks to Northern Hemisphere–based librarians at the Scott Polar Research Institute, Kew Gardens, and the University of Illinois, and to the captain and crew of the *Ortelius* for the journey south. Jim Kennett and Jim Zachos offered illuminating recollections of their experiences with the Ocean Drilling Program in the Antarctic, while Ingrid Gnerlich and colleagues at Princeton University Press maintained the highest possible standards and were unfailingly encouraging. My special thanks to Dennis Sears for his patient expertise with images, and to the three anonymous reviewers of the manuscript for their keen suggestions and corrections. Gratitude, lastly, is due my family, who endured both my intermittent absences and, worse, the daily absentmindedness of the mental traveler to the ice.

BIBLIOGRAPHY

General Reference

Day, David. *Antarctica: A Biography.* Oxford: Oxford University Press, 2013.

Florindo, Fabio, and Martin Siegert. *Antarctic Climate Evolution.* Amsterdam: Elsevier, 2009.

Fogg, G. E. *A History of Antarctic Science.* Cambridge: Cambridge University Press, 1992.

King, J. C., and J. Turner. *Antarctic Meteorology and Climatology.* Cambridge: Cambridge University Press, 1997.

Knox, George A. *Biology of the Southern Ocean.* 2nd ed. Boca Raton: CRC Press, 2007.

Menzies, John (ed.). *Modern and Past Glacial Environments.* Oxford: Butterworth Heinemann, 2002.

Mill, Hugh Robert. *The Siege of the South Pole.* London: A. Rivers, 1905.

Prothero, Donald R., and William A. Berggren. *Eocene-Oligocene Climatic and Biotic Evolution.* Princeton: Princeton University Press, 1992.

Riffenburgh, Beau (ed.). *Encyclopaedia of the Antarctic.* New York: Routledge, 2007.

Summerhayes, Colin P. *Earth's Climate Evolution.* Chichester: Wiley-Blackwell, 2015.

Thomas, David N. *Sea Ice.* 3rd ed. Hoboken, NJ: Wiley & Sons, 2017.

Walton, David, W. H. (ed). *Antarctica: Global Science from a Frozen Continent.* Cambridge: Cambridge University Press, 2013.

Williams, Tony D. *The Penguins.* Oxford: Oxford University Press, 1995.

Primary Sources

Amundsen, Roald. *The South Pole: An Account of the Norwegian Antarctic Expedition in the Fram, 1912–13*. Trans. A. G. Chater. London: J. Murray, 1913.

British Antarctic Expedition, 1839–43. Letters and Journals. Public Records Office, Kew.

Darwin, Charles. *Journal of Researches into the Geology and Natural History of the Various Countries Visited by HMS Beagle*. London: H. Colburn, 1839.

D'Urville, Jules-Sébastien Dumont. *Voyage au Pole Sud et dans L'Océanie sur les Corvettes L'Astrolabe et La Zélée. Histoire du Voyage*. 10 vols. Paris: Gide, 1842–45.

Edwards, Philip. *Last Voyages: Cavendish, Hudson, Ralegh; The Original Narratives*. Oxford: Clarendon, 1988.

Erskine, Charles. *Twenty Years before the Mast*. Philadelphia: George W. Jacobs, 1896.

Hooker, Joseph. Papers. Kew Gardens Library.

———. *The Botany of the Antarctic Voyage of H. M. Discovery Ships* Erebus *and* Terror *in the Years 1839–1843*. 4 vols. London: Reeve Bros., 1847–60.

Huxley, Leonard. *The Life and Letters of Sir Joseph Dalton Hooker*. 2 vols. New York: Appleton, 1918.

Mawson, Douglas. *The Home of the Blizzard: A True Story of Antarctic Survival*. New York: St. Martin's, 1998.

Nordenskjöld, Otto, and Gunnar Andersson. *Antarctica; or, Two Years amongst the Ice of the South Pole*. New York: Macmillan, 1905.

Palmer, J. C. *Thulia: A Tale of the Antarctic*. New York: S. Colman, 1843.

Reynolds, Jeremiah. *Address on the Subject of a Surveying and Exploring Expedition to the Pacific Ocean and South Seas*. New York: Harper & Bros., 1836.

Reynolds, William. *The Private Journal of William Reynolds: United States Exploring Expedition, 1838–1842*, ed. Nathaniel Philbrick and Thomas Philbrick. New York: Penguin, 2004.

Ross, James Clark. *A Voyage of Discovery and Research in the Southern and Antarctic Regions During the Years 1839–43*. 2 vols. London: J. Murray, 1847.

Shackleton, Ernest. *The Heart of the Antarctic*. 2 vols. Philadelphia: Lippincott, 1909.

Stokes, Pringle. "The Journal of HMS *Beagle* in the Strait of Magellan [1827]." In *Four Travel Journals: The Americas, Antarctica and Africa, 1775–1874*, ed. Herbert K. Beals, et al. London: Hakluyt Society, 2007, pp. 141–252.

"Visite de M. l'amiral Duperré, ministre de la marine, au Muséum." *Annales Maritimes et Coloniales* 2.2 (1841): 101–3.

Weddell, James. *A Voyage towards the South Pole: Performed in the Years 1822–24*. London: Longman, 1825.

Wilkes, Charles. *Autobiography*, ed. William James Morgan et al. Washington, DC: Department of the Navy, 1978.

———. *Narrative of the United States Exploring Expedition*, 5 vols. Philadelphia: Lea and Blanchard, 1845.

Secondary Sources

Bartlett, Harley Harris. "The Report of the Wilkes Expedition, and the Work of the Specialists in Science." *Proceedings of the American Philosophical Society* 82.5 (1940): 601–705.

Cawood, John. "The Magnetic Crusade: Science and Politics in Early Victorian Britain." *Isis* 70.254 (1979): 493–518.

———. "Terrestrial Magnetism and the Development of International Collaboration in the Early Nineteenth Century." *Annals of Science* 34 (1977): 551–87.

Cohen, Morton. *Lewis Carroll: A Biography*. New York: Vintage, 1996.

Delépine, Gracie. *Les Iles Australes Françaises*. Rennes: Éditions Ouest-France, 1995.

Duyker, Edward. *Dumont D'Urville: Explorer and Polymath*. Honolulu: University of Hawaii Press, 2014.

Fleming, James Rodger. *Meteorology in America, 1800–1870*. Baltimore: Johns Hopkins University Press, 1990.

Goodell, Jeff. "The Doomsday Glacier." *Rolling Stone* 1287 (May 18, 2017): 44–51.

Malin, S. R. C., and D. R. Barraclough. "Humboldt and the Earth's Magnetic Field." *Quarterly Journal of the Royal Astronomical Society* 32 (1991): 279–93.

Mawer, Granville Allen. *South by Northwest: The Magnetic Crusade and the Contest for Antarctica*. Adelaide: Wakefield, 2006.

McEwan, Colin, Luis A. Borrero, and Alfred Prieto. *Patagonia: Natural History, Prehistory, and Ethnography at the Uttermost Ends of the Earth*. London: British Museum, 1997.

Moss, Chris. *Patagonia: A Cultural History*. Oxford: Oxford University Press, 2008.

Philbrick, Nathaniel. *Sea of Glory: America's Voyage of Discovery; The U.S. Exploring Expedition, 1838–1842*. New York: Viking, 2003.

Riffenburgh, Beau. *Shackleton's Forgotten Expedition: The Voyage of the Nimrod*. New York: Bloomsbury, 2004.

Ross, M. J. *Polar Pioneers: John Ross and James Clark Ross*. Montreal: McGill-Queen's University Press, 1994.

———. *Ross in the Antarctic*. Whitby: Caedmon, 1982.

Stanton, William. *The Great United States Exploring Expedition of 1838–1842*. Berkeley: University of California Press, 1975.

Viola, Herman, and Carolyn Margolis. *Magnificent Voyagers: The U.S. Exploring Expedition, 1838–42*. Washington, DC: Smithsonian, 1985.

Scientific Sources

Acosta Hospitaleche, Carolina. "New Giant Penguin Bones from Antarctica: Systematic and Paleobiological Significance." *Comptes Rendus Palevol* 13 (2014): 555–60.

Acosta Hospitaleche, Carolina, Marcelo Reguero, and Alejo Scarano. "Main Pathways in the Evolution of the Paleogene Antarctic *Sphenisciformes*." *Journal of South American Earth Sciences* 43 (2013): 101–11.

Alley, Richard B., et al. "Oceanic Forcing of Ice-Sheet Retreat: West Antarctica and More." *Annual Review of Earth and Planetary Sciences* 43 (2015): 207–31.

Baker, Allan, et al. "Multiple Gene Evidence for Expansion of Extant Penguins out of Antarctica Due to Global Cooling." *Proceedings of the Royal Society* B.273 (2006): 11–17.

Barker, Peter. "A History of Antarctic Cenozoic Glaciation: View from the Margin." In F. Florindo and M. Siegert (eds.), *Antarctic Climate Evolution*. Developments in Earth and Environmental Sciences 8. Amsterdam: Elsevier, 2009, pp. 33–83.

Barker, Peter, et al. "Onset and Role of the Antarctic Circumpolar Current." *Deep-Sea Research II* 54 (2007): 2388–98.

Berger, W. H. "Cenozoic Cooling, Antarctic Nutrient Pump, and the Evolution of Whales." *Deep-Sea Research II* 54 (2007): 2399–2421.

Borrero, Luis A. "Human Dispersal and Climatic Conditions during Late Pleistocene Times in Fuego-Patagonia." *Quaternary International* 53–54 (1999): 93–99.

Borrero, Luis A., and Nora V. Franco. "Early Patagonian Hunter-Gatherers: Subsistence and Technology." *Journal of Anthropological Research* 53 (1997): 219–39.

Breza, J. R., and S. W. Wise Jr. "Lower Oligocene Ice-Rafted Debris on the Kerguelen Plateau: Evidence for East Antarctic Continental Glaciation." In S. W. Wise Jr., R. Schlich, et al. (eds.), *Proceedings of the Ocean Drilling Program, Scientific Results* 120. Washington, DC: Integrated Ocean Drilling Program Management International, 1992, pp. 161–78.

Briones, Claudia, and José L. Lanata (eds.). *Archaeological and Anthropological Perspectives on the Native Peoples of Pampa, Patagonia, and Tierra del Fuego to the Nineteenth Century*. Westport, CT: Bergin & Harvey, 2002.

Carbulotto, A. "New Insights into Quaternary Glacial Dynamic Changes on the George V Land Continental Margin (East Antarctica)." *Quaternary Science Reviews* 25 (2006): 3029–49.

Case, Judd A. "Evidence from Fossil Vertebrates for a Rich Eocene Antarctic Marine Environment." *Antarctic Research Series* 56 (1992): 119–30.

———. "Paleogene Floras from Seymour Island, Antarctic Peninsula." *Geological Society of America* 169 (1988): 523–39.

Chapman, V. J. *Seaweeds and Their Uses.* 2nd ed. London: Methuen, 1970.

Chaput, J., et al. "Near-Surface Environmentally Forced Changes in the Ross Ice Shelf Observed with Ambient Seismic Noise." *Geophysical Research Letters* 45 (Oct. 16, 2018): 11,187–96.

Clarke, Julia A., et al. "Fossil Evidence for Evolution of the Shape and Color of Penguin Feathers." *Science* 330 (Nov. 12, 2010): 954–57.

———. "Paleogene Equatorial Penguins Challenge the Proposed Relationship between Biogeography, Diversity, and Cenozoic Climate Change." *Proceedings of the National Academy of Sciences* 104.28 (July 10, 2007): 11545–50.

Coffin, Millard F., et al. "Kerguelen Hotspot Magma Output since 130 Ma." *Journal of Petrology* 43.7 (2002): 1121–39.

Contreras, Lineth, et al. "Early to Middle Eocene Vegetation Dynamics at the Wilkes Land Margin (Antarctica)." *Review of Palaeobotany and Palynology* 197 (2013): 119–42.

Cook, Carys, et al. "Glacial Erosion of East Antarctica in the Pliocene: A Comparative Study of Multiple Marine Sediment Provenance Tracers." *Chemical Geology* 466 (2017): 199–218.

———. "Sea Surface Temperature Control on the Distribution of Far-Traveled Southern Ocean Ice-Rafted Detritus during the Pliocene." *Paleoceanography* 29 (2014): 533–48.

Coronato, A., et al. "Palaeoenvironmental Conditions during the Early Peopling of Southernmost South America (Late Glacial—Early Holocene, 14–8ka B.P.)." *Quaternary International* 53–54 (2002): 77–92.

DeConto, Robert M., and David Pollard. "Contribution of Antarctica to Past and Future Sea-Level Rise." *Nature* 531 (2016): 591–97.

Deschamps, Pierre, et al. "Ice-Sheet Collapse and Sea-Level Rise at the Bølling Warming 14,600 Years Ago." *Nature* 483 (2012): 559–64.

Dillehay, Tom D., et al. "New Archaeological Evidence for an Early Human Presence at Monte Verde, Chile." *Public Library of Science* 10.12 (2015): e0141923.

Ding, Qinghua, et al. "Winter Warming in West Antarctica Caused by Central Tropical Pacific Warming." *Nature Geoscience* 4 (2011): 398–403.

Drewry, David J. "Radio Echo Sounding Map of Antarctica." *Polar Record* 17.109 (1975): 359–74.

Ehrmann, W. U., M. J. Hambrey, J. G. Baldauf, J. Barron, B. Larsen, A. Mackensen, S. W. Wise Jr., and J. C. Zachos. "History of Antarctic Glaciation: An Indian Ocean Perspective." *Synthesis of Results from Scientific Drilling in the Indian Ocean*. Geophysical Monograph 70. Washington, DC: American Geophysical Union, 1992.

Eisen, O., and C. Kottmeier. "On the Importance of Leads in Sea Ice to the Energy Balance and Ice Formation in the Weddell Sea." *Journal of Geophysical Research* 105 (2000): 14,045–60.

Erlandson, Jon M., et al. "Ecology of the Kelp Highway: Did Marine Resources Facilitate Human Dispersal from Northeast Asia to the Americas?" *Journal of Island & Coastal Archaeology* 10 (2015): 392–411.

———. "How Old Is MVII? Seaweeds, Shorelines, and the Pre-Clovis Chronology at Monte Verde, Chile." *Journal of Island & Coastal Archaeology* 3 (2008): 277–81.

———. "The Kelp Highway Hypothesis: Marine Ecology, the Coastal Migration Theory, and the Peopling of the Americas." *Journal of Island & Coastal Archaeology* 2 (2007): 161–74.

Escutia, Carlota, Henk Brinkhuis, and the Expedition 318 Scientists. "From Greenhouse to Icehouse at the Wilkes Land Antarctic Margin: IODP Expedition 318 Synthesis of Results." Ruediger Stein, Donna K. Blackman, Fumio Inagaki, and Hans-Christian Larsen (eds.), *Developments in Marine Geology* 7 (2014): 295–328.

Escutia, Carlota, Henk Brinkhuis, A. Klaus, and the Expedition 318 Scientists. *Proceedings of the Integrated Ocean Drilling Program* 318. Washington, DC: Integrated Ocean Drilling Program Management International, 2011.

Evans, Michael E., and Friedrich Heller. *Environmental Magnetism: Principles and Applications of Environmagnetics*. San Diego: Academic, 2003.

Falkowski, Paul G. "The Evolution of Modern Eukaryotic Phytoplankton." *Science* 305 (2004): 354–60.

Feldmann Johannes, and Anders Levermann. "Collapse of the West Antarctic Ice Sheet after Local Destabilization of the Amundsen

Basin." *Proceedings of the National Academy of Science* 112.46 (2015): 14191–96.

Fordyce, Ewan. "Whale Evolution and Oligocene Southern Ocean Environments." *Palaeogeography, Palaeoclimatology, Palaeoecology* 31 (1980): 319–36.

Francis, J. E., D. Pirrie, and J. A. Crame (eds). *Cretaceous-Tertiary High-Latitude Palaeoenvironments: James Ross Basin, Antarctica.* London: Geological Society, 2006.

Frankel, Henry R. *The Continental Drift Controversy: Paleomagnetism and Confirmation of Drift.* Cambridge: Cambridge University Press, 2012.

Fraser, Ceridwen I. 2009. "Kelp Genes Reveal Effects of Sub-Antarctic Sea Ice during Last Glacial Maximum." *Proceedings of the National Academy of Sciences* 106.9 (2009): 3249–53.

Goebel, Ted, et al. "The Late Pleistocene Dispersal of Modern Humans in the Americas." *Science* 319 (2008): 1497–1502.

Gordon, Arnold L. "Western Weddell Sea Thermohaline Stratification," in *Ocean, Ice, and Atmosphere: Interactions at the Antarctic Continental Margin*, ed. Stanley Jacobs and Ray Weiss. Washington, DC: 1998, pp. 215–40.

Gordon, Arnold L., et al. "Deep and Bottom Water of the Weddell Sea's Western Rim." *Science* 262 (1993): 95–97.

Graf, Kelly E., Caroline V. Ketron, and Michael R. Waters (eds.). *Paleoamerican Odyssey.* College Station: Texas A&M University Press, 2014.

Graham, Michael H., et al. "Ice Ages and Ecological Transitions on Temperate Coasts." *Trends in Ecology and Evolution* 18.1 (2003): 33–40.

Griffiths, D. H., and R. F. King. "Natural Magnetization of Igneous and Sedimentary Rocks." *Nature* 173 (June 12, 1954): 1114–17.

Hansen, Melissa A. and Sandra Passchier. "Oceanic Circulation Changes during Early Pliocene Marine Ice-Sheet Instability in Wilkes Land, East Antarctica." *Geo-Marine Letters* 37 (2017): 207–13.

Hauer, Matthew E. "Migration Induced by Sea-Level Rise Could Reshape the U.S. Population Landscape." *Nature Climate Change* 7 (2017): 321–25.

Heimann, A., et al. "A Short Interval of Jurassic Continental Flood Basalt Volcanism in Antarctica as Demonstrated by 40Ar/39Ar

Geochronology." *Earth and Planetary Science Letters* 121 (1994): 19–41.

Hellegatte, Stephane, et al. "Future Flood Losses in Major Coastal Cities." *Nature Climate Change* 3 (2013): 802–6.

Hernandez, Miquel, et al. "Fuegian Cranial Morphology: The Adaptation to a Cold, Harsh Environment." *American Journal of Physical Anthropology* 103 (1997): 103–17.

Initial Reports of the Deep Sea Drilling Project, vol. 21. Washington, DC: National Science Foundation, 1973.

Initial Reports of the Deep Sea Drilling Project, vol. 29. Washington, DC: National Science Foundation, 1975.

Irving, E. "Palaeomagnetic and Palaeoclimatological Aspects of Polar Wandering." *Geofisica Pura e Applicata* 33.1 (1956): 23–41.

Jablonski, Nina G. (ed.). *The First Americans: The Pleistocene Colonization of the New World*. San Francisco: California Academy of Sciences, 2002.

Jadwiszczak, Piotr. "Partial Limb Skeleton of a "Giant Penguin" *Anthropornis* from the Eocene of Antarctic Peninsula." *Polist Polar Research* 33.3 (2012): 259–74.

———. "Penguin Past: The Current State of Knowledge." *Polish Polar Research* 30.1 (2009): 3–28.

Jadwiszczak, Piotr, and Thomas Mors. "Aspects of Diversity in Early Antarctic Penguins." *Acta Palaeontologica Polonica* 56.2 (2011): 269–77.

Jamieson, S.S.R., and D. E. Sugden. "Landscape Evolution of Antarctica." In *Antarctica: A Keystone in a Changing World*, ed. A. K. Cooper, et al. Washington, DC: National Academies Press, 2008, p.39–54.

Jevrejeva, Svetlana, et al. "Coastal Sea Level Rise with Warming above 2°C." *Proceedings of the National Academy of Science* 113.47 (Nov. 22, 2016): 13,342–47.

Jones, David A., and Ian Simmonds. "A Climatology of Southern Hemisphere Extra-tropical Cyclones." *Climate Dynamics* 9 (1993): 131–45.

Jordan, Richard W. and Catherine E. Stickley. "Diatoms as Indicators of Paleoceanographic Events." In John P. Smol and Eugene F. Stoermer (eds.), *The Diatoms: Applications for the Environmental and Earth Sciences*, 2nd ed. Cambridge: Cambridge University Press, 2010, pp. 424–53.

Jordan, T. A. "Hypothesis for Mega-Outburst Flooding from a Palaeo-Subglacial Lake beneath the East Antarctic Ice Sheet." *Terra Nova* 22 (2010): 283–89.

Joughin, Ian, et al. "Marine Ice Sheet Collapse Potentially Under Way for the Thwaites Glacier Basin, West Antarctica." *Science* 344 (2014): 735–38.

Kennett, James P. "Cenozoic Evolution of Antarctic Glaciation, the Circum-Antarctic Ocean, and Their Impact on Global Paleoceanography." *Journal of Geophysical Research* 82.27 (1977): 3843–60.

———. "Recognition and Correlation of the Kapitean Stage (Upper Miocene, New Zealand)." *New Zealand Journal of Geology and Geophysics* 10 (1967): 1051–63.

Kennett, James P., and Stanley V. Margolis. "Antarctic Glaciation during the Tertiary Recorded in Sub-Antarctic Deep-Sea Cores." *Science* 170.3962 (1970): 1085–87.

Kennett, James P., and Nicholas Shackleton. "Oxygen Isotopic Evidence for the Development of the Psychrosphere 38Myr ago." *Nature* 260 (1976): 513–15.

Kennett, James P., et al. "Australian-Antarctic Continental Drift, Palaeocirculation Changes and Oligocene Deep-Sea Erosion." *Nature: Physical Science* 239 (1972): 51–5.

———. "Development of the Circum-Antarctic Current." *Science* 186.4159 (1974): 144–47.

Khazendar, Ala, et al. "Rapid Submarine Ice Melting in the Grounding Zones of Ice Shelves in West Antarctica." *Nature Communications* 7 (Oct. 25, 2016).

Ksepka, Daniel T., and Tatsuro Ando. "Penguins Past, Present, and Future: Trends in the Evolution of the *Sphenisciformes*." In Gareth Dyke and Gary Kaiser (eds.), *Living Dinosaurs: The Evolutionary History of Modern Birds*. New York: John Wiley and Sons, 2011, pp. 155–86.

Ksepka, Daniel T., Sara Bertelli, and Norberto P. Giannini. "The Phylogeny of the Living and Fossil *Sphenisciformes* (Penguins)." *Cladistics* 22 (2006): 412–41.

Lamb, H. H. " The Southern Westerlies: A Preliminary Survey, Main Characteristics, and Apparent Associations." *Quarterly Journal of the Royal Meteorological Society* 85.363 (1959): 1–23.

Livermore, Roy, et al. "Drake Passage and Cenozoic Climate: An Open and Shut Case?" *Geochemistry, Geophysics, Geosystems* 8.1 (2007): Q01005.

Loewe, F. "The Land of Storms." *Weather* 27.3 (1972): 110–12.

Madsen, D. B. (ed.). *Entering America: Northeast Asia and Beringia before the Last Glacial Maximum.* Salt Lake City: University of Utah Press, 2004.

Maher, Barbara A. "Environmental Magnetism and Climate Change." *Contemporary Physics* 48.5 (2007): 247–74.

Marsh, Oliver J., et al. "High Basal Melting Forming a Channel at the Grounding Line of Ross Ice Shelf, Antarctica." *Geophysical Research Letters* 43 (Jan. 14, 2016).

Mengel, M., and A. Levermann. "Ice Plug Prevents Irreversible Discharge from East Antarctica." *Nature Climate Change* 4 (June 2014): 451–55.

Milillo, P., et al. "Heterogenous Retreat and Ice Melt of Thwaites Glacier, West Antarctica." *Science Advances* 5.1 (2019).

Miller, Eric R. "American Pioneers in Meteorology." *Monthly Weather Review* (1933) 61.7 (1933): 189–93.

Miller, Kenneth G. "Climate Threshold at the Eocene-Oligocene Transition: Antarctic Ice Sheet Influence on Ocean Circulation." In C. Koeberl and A. Montanari (eds.), *The Late Eocene Earth: Hothouse, Icehouse, and Impacts.* GSA Special Papers 452. Boulder, CO: Geological Society of America, 2009, pp. 169–78.

Mouginot, J., et al. "Sustained Increase in Ice Discharge from the Amundsen Sea Embayment, West Antarctica, from 1973 to 2013." *Geophysical Research Letters* 41 (2014): 1576–84.

Naish, T., et al. "Obliquity-Paced Pliocene West Antarctic Ice Sheet Oscillations." *Nature* 458 (2009): 322–29.

Nicholls, Robert J., et al. "Sea-Level Rise and Its Possible Impacts Given a 'Beyond 4°C World' in the Twenty-First Century." *Philosophical Transactions of the Royal Society* 369 (2011): 161–81.

Nicolaysen, K., et al. "40Ar/39Ar Geochronology of Flood Basalts from the Kerguelen Archipelago, Southern Indian Ocean: Implications for Cenozoic Eruption Rates of the Kerguelen Plume." *Earth and Planetary Science Letters* 174 (2000): 313–28.

Noback, Marlijn L., et al. "Climate-Related Variation of the Human Nasal Cavity." *American Journal of Physical Anthropology* 145 (2011): 599–614.

Ohneiser, C., et al. "Characterisation of Magnetic Minerals from Southern Victoria Land, Antarctica." *New Zealand Journal of Geology and Geophysics* 58.1 (2015): 52–65.

Parish, Thomas. "The Katabatic Winds of Cape Denison and Port Martin." *Polar Record* 20.129 (1981): 525–32.

———. 1984. "A Numerical Study of Strong Katabatic Winds over Antarctica." *Monthly Weather Review* 112 (1984): 545–54.

———. "On the Interaction between Antarctic Katabatic Winds and Tropospheric Motions in the High Southern Latitudes." *Australian Meteorological Magazine* 40 (1992): 149–67.

———. 1987. "The Surface Windfield over the Antarctic Ice Sheets." *Nature* 328 (1987): 51–4.

Parish, Thomas, and David H. Bromwich. "Re-examination of the Near Surface Airflow over the Antarctic Continent and Implications on Atmospheric Circulations at High Southern Latitudes." *Monthly Weather Review* 135 (2007): 1961–73.

Parish, Thomas, and Richard Walker. "A Re-examination of the Winds of Adélie Land, Antarctica." *Australian Meteorological Magazine* 55 (2006): 105–17.

Patterson, M. O., et al. "Orbital Forcing of the East Antarctic Ice Sheet during the Pliocene and Early Pleistocene." *Nature Geoscience* 7 (Nov. 2014): 841–47.

Pennycuick, C. J. "The Flight of Petrels and Albatrosses (*Procellariiformes*), Observed in South Georgia and Its Vicinity." *Philosophical Transactions of the Royal Society of London* 300 (1982): 75–106.

Piana, Ernesto L., and Luis A. Orquera. "The Southern Top of the World: The First Peopling of Patagonia and Tierra del Fuego and the Cultural Endurance of the Fuegian Sea-Nomads." *Arctic Anthropology* 46.1–2 (2009): 103–17.

Pierce, Elizabeth L. "Evidence for a Dynamic East Antarctic Ice Sheet during the Mid-Miocene Climate Transition." *Earth and Planetary Science Letters* 478 (2017): 1–13.

Pitulko, V. V., et al. "The Yana RHS Site: Humans in the Arctic before the Last Glacial Maximum." *Science* 303 (2004): 52–56.

Pollock, David E. "The Role of Diatoms, Dissolved Silicate and Antarctic Glaciation in Glacial/Interglacial Climatic Change: A Hypothesis." *Global and Planetary Change* 14 (1997): 113–25.

Pritchard, H. D., et al. "Antarctic Ice-Sheet Loss Driven by Basal Melting of Ice Shelves." *Nature* 484 (2012): 502–5.

Redfield, William C. "Remarks on the Prevailing Storms of the Atlantic Coast, of the North American States." *American Journal of Science and Arts* 20.1 (1831): 17–53.

Reid, William. *An Attempt to Develop the Law of Storms by Means of Facts*. London: J. Weale, 1838.

Reinardy, B.T.I., et al. "Repeated Advance and Retreat of the East Antarctic Ice Sheet on the Continental Shelf during the Early Pliocene Warm Period." *Palaeogeography, Palaeoclimatology, Palaeoecology* 422 (2015): 65–84.

Rignot, Eric, et al. "Widespread, Rapid Grounding Line Retreat of Pine Island, Thwaites, Smith, and Kohler Glaciers, West Antarctica, from 1992 to 2011." *Geophysical Research Letters* 41 (2014): 3502–9.

Roberts, Andrew, et al. "Environmental Magnetic Record of Paleoclimate, Unroofing of the Transantarctic Mountains, and Volcanism in Late Eocene to Early Miocene Glaci-Marine Sediments from the Victoria Land Basin, Ross Sea, Antarctica." *Journal of Geophysical Research: Solid Earth* 118 (2013): 1845–61.

Runcorn, S. K. "Climatic Change through Geological Time in the Light of the Palaeomagnetic Evidence for Polar Wandering and Continental Drift." *Royal Meteorological Society Quarterly Journal* 87.373 (1961): 282–313.

Ruskin, John. "Remarks on the Present State of Meteorological Science." *Transactions of the Meteorological Society* 1 (1839): 56–59.

Sabine, Edward. *An Account of Experiments to Determine the Figure of the Earth*. London: J. Murray, 1825.

———. "On Periodical Laws Discoverable in the Mean Effects of the Larger Magnetic Disturbances." Pts. 1 and 2. *Philosophical Transactions of the Royal Society of London* 141 (1851): 123–39; 142 (1852): 103–24.

———. "On What the Colonial Magnetic Observatories Have Accomplished." *Proceedings of the Royal Society of London* 8 (1856–57): 396–413.

———. "Report on the Variations of the Magnetic Intensity Observed at Different Points on the Earth's Surface." *Report of the Seventh Meeting of the British Association for the Advancement of Science, 1837*. London: J. Murray, 1838.

Sagnotti, Leonardo, et al. "Environmental Magnetic Record of Antarctic Palaeoclimate from Eocene/Oligocene Glaciomarine Sediments, Victoria Land Basin." *Geophysical Journal International* 134 (1998): 653–62.

———. "Environmental Magnetic Record of the Eocene-Oligocene Transition in CRP-3 Drillcore, Victoria Land Basin, Antarctica." *Terra Antartica* 8.4 (2001): 507–16.

Sangiorgi, Francesca, et al. "Southern Ocean Warming and Wilkes Land Ice Sheet Retreat during the Mid-Miocene." *Nature Communications* 9 (2018).

Scott, Ryan C., et al. "Meteorological Drivers and Large-Scale Climate Forcing of West Antarctic Surface Melt." *Journal of Climate* 32.3 (2019): 665–84.

Simpson, G. G. "Review of Fossil Penguins from Seymour Island." *Proceedings of the Royal Society of London* B.178 (1971): 357–87.

Spear, Larry B., and David G. Ainley. "Flight Behaviour of Seabirds in Relation to Wind Direction and Wing Morphology." *Ibis* 139 (1997): 221–33.

Stickley, Catherine, et al. "Timing and Nature of the Deepening of the Tasmanian Gateway." *Paleoceanography* 19 (2004): PA4027.

Stonehouse, Bernard. *The Biology of Penguins*. London: Macmillan, 1975.

Sugden, D. E., et al. "Late-Glacial Glacier Events in Southernmost South America: A Blend of 'Northern' and 'Southern' Hemispheric Climatic Signals?" *Geografiska Annaler* 87A (2005): 273–88.

Thomas, Daniel B., Daniel T. Ksepka, and R. Ewan Fordyce. "Penguin Heat-Retention Structures Evolved in a Greenhouse Earth." *Biology Letters* 7 (2011): 461–64.

Timmermann, R., et al. "The Role of Sea Ice in the Fresh-Water Budget of the Weddell Sea, Antarctica." *Annals of Glaciology* 33 (2001): 419–24.

Vine, F. J., and D. H. Matthews. "Magnetic Anomalies over Oceanic Ridges." *Nature* 4897 (Sept. 7, 1963): 947–49.

Wang, Sijia, et al. "Genetic Variation and Population Structure in Native Americans." *PLoS Genetics* 3.11 (2007): 2049–67.

Wei, Wuchang, et al. "Paleoceanographic Implications of Eocene-Oligocene Calcareous Nannofossils from Sites 711 and 748 in the Indian Ocean." In S. W. Wise Jr. and R. Schlich, et al. (eds.), *Proceedings of the Ocean Drilling Program, Scientific Results* 120 (1992): 979–99.

Wendler, Gerd, et al. "On the Extraordinary Katabatic Winds of Adélie Land." *Journal of Geophysical Research* 102 (1997): 4463–74.

Wilson, Gary S. "Magnetobiostratigraphic Chronology of the Eocene-Oligocene Transition in the CIROS-1 Core, Victoria Land Margin, Antarctica: Implications for Antarctic Glacial History." *Geological Society of America Bulletin* 110.1 (1998): 35–47.

Winkelmann, Ricarda, et al. "Combustion of Available Fossil Fuel Resources Sufficient to Eliminate the Antarctic Ice Sheet." *Science Advances* 1.8 (Sept. 4, 2015).

Wise, Sherwood W., Jr., James R. Breza, David M. Harwood, and Wuchang Wei. "Paleogene Glacial History of Antarctica." In D. W. Muller, J. A. McKenzie, and H. Weissert (eds.), *Controversies in Modern Geology: Evolution of Geological Theories in Sedimentology, Earth History, and Tectonics*. London: Academic, 1991, pp. 133–72.

Woodburne, Michael O., and Judd Case. "Dispersal, Vicariance, and the Late Cretaceous to Early Tertiary Land Mammal Biogeography from South America to Australia." *Journal of Mammalian Evolution* 3.2 (1996): 121–61.

Young, Duncan A., et al. "A Dynamic Early East Antarctic Ice Sheet Suggested by Ice-Covered Fjord Landscapes." *Nature* 474 (June 2, 2011): 72–75.

Zachos, James C., William A. Berggren, Marie-Pierre Aubry, and Andreas Mackensen. "Isotope and Trace Element Geochemistry of

Eocene and Oligocene Foraminifers from Site 748, Kerguelen Plateau." In S. W. Wise Jr. and R. Schlich, et al. (eds.), *Proceedings of the Ocean Drilling Program, Scientific Results* 120. Washington, DC: Integrated Ocean Drilling Program Management International, 2011, pp. 839–54.

Zachos, James C., James R. Breza, Sherwood W. Wise. "Early Oligocene Ice-Sheet Expansion on Antarctica: Stable Isotope and Sedimentological Evidence from Kerguelen Plateau, Southern Indian Ocean." *Geology* 20 (1992): 569–73.

Zachos, James C., et al. "Abrupt Climate Change and Transient Climates during the Paleogene: A Marine Perspective." *Journal of Geology* 101 (1993): 191–213.

Zinsmeister, William J. "Biogeographic Significance of the Late Mesozoic and Early Tertiary Molluscan Faunas of Seymour Island (Antarctic Peninsula) to the Final Breakup of Gondwanaland." In *Historical Biogeography, Plate Tectonics, and the Changing Environment,* ed. Jane Gray and Arthur J. Boucot. Corvallis: Oregon State University Press, 1976, pp. 349–55.

———. "Early Geological Exploration of Seymour Island, Antarctica." *Geological Society of America* 169 (1988): 1–16.

INDEX